Solar Photovoltaic Systems Installer

Trainee Guide

Pearson

Boston Columbus Indianapolis New York San Francisco Upper Saddle River
Amsterdam Cape Town Dubai London Madrid Milan Munich Paris Montreal Toronto
Delhi Mexico City São Paulo Sydney Hong Kong Seoul Singapore Taipei Tokyo

NCCER
President: Don Whyte
Director of Product Development: Daniele Stacey
Solar Photovoltaic Systems Installer Project Manager: Matt Tischler
Production Manager: Tim Davis

Quality Assurance Coordinator: Debie Ness
Editor: Chris Wilson
Desktop Publishing Coordinator: James McKay
Production Specialist: Laura Wright

Writing and development services provided by Topaz Publications, Liverpool, NY
Lead Writer/Project Manager: Veronica Westfall
Desktop Publisher: Joanne Hart
Art Director: Megan Paye

Permissions Editors: Andrea LaBarge, Alison Richmond
Writers: Roy Parker, Troy Staton, Pat Vidler

Pearson Education, Inc.
Editorial Director: Vernon R. Anthony
Senior Product Manager: Lori Cowen
Senior Managing Editor: JoEllen Gohr
Senior Project Manager: Steve Robb
Operations Supervisor: Deidra M. Skahill
Art Director: Jayne Conte
Cover photo: istockphoto.com
Director of Marketing: David Gessell
Executive Marketing Manager: Derril Trakalo

Senior Marketing Coordinator: Alicia Wozniak
Marketing Assistant: Les Roberts
Full-Service Project Management: Michael B. Kopf, S4Carlisle
    Publishing Services
Composition: S4Carlisle Publishing Services
Printer/Binder: LSC Communications
Cover Printer: LSC Communications
Text Fonts: Palatino and Univers

Credits and acknowledgments for content borrowed from other sources and reproduced, with permission, in this textbook appear at the end of each module.

This information is general in nature and intended for training purposes only. Actual performance of activities described in this manual requires compliance with all applicable operating, service, maintenance, and safety procedures under the direction of qualified personnel. References in this manual to patented or proprietary devices do not constitute a recommendation of their use.

15 2022

PEARSON

ISBN-13: 978-0-13-257110-4
ISBN-10:   0-13-257110-2

Preface

To the Trainee

The solar power industry is a rapidly growing field that is expected to help ease human dependence on the use of fossil fuels. Solar panels are now rated to produce up to 600 volts of electricity, and the cost of purchasing and installing these panels for both residential and commercial purposes has been reduced considerably. Because of this, the need for solar photovoltaic installers has increased and is projected to grow with the demand for solar installations.

Through both government and private initiatives, the increased need for qualified solar photovoltaic installers will provide you with an opportunity to join an expanding and viable workforce. According to the Energy Information Administration, domestic shipments of solar panels have increased by nearly 2,400 percent since 2000, and the U.S. Department of Labor expects the solar photovoltaic workforce to grow by 50 percent in the coming years.

This textbook covers specific information, equipment, and installation techniques that are valuable to photovoltaic systems installers, including design considerations, system components, site analysis, troubleshooting, and maintenance. The *Solar Photovoltaic Systems Installer* curriculum is designed to assist trainees in obtaining the North American Board of Certified Energy Practitioners (NABCEP) PV Installer Certification.

By taking this course, you'll be equipped with the knowledge necessary to be a successful member of the solar energy workforce. Today, the growth of power generation is tied to clean and renewable energy. The skills you learn in this textbook will help provide you with a lifetime of opportunity in this field.

We wish you success as you progress through this training program. Should you have any comments on how NCCER might improve upon this textbook, please complete the User Update form located at the back of each module and send it to us. We will always consider and respond to input from our customers.

We invite you to visit the NCCER website at **www.nccer.org** for information on the latest product releases and training, as well as online versions of the *Cornerstone* magazine and Pearson's product catalog.

Your feedback is welcome. You may email your comments to **curriculum@nccer.org** or send general comments and inquiries to **info@nccer.org**.

NCCER Standardized Curriculum

NCCER is a not-for-profit 501(c)(3) education foundation established in 1996 by the world's largest and most progressive construction companies and national construction associations. It was founded to address the severe workforce shortage facing the industry and to develop a standardized training process and curricula. Today, NCCER is supported by hundreds of leading construction and maintenance companies, manufacturers, and national associations. The NCCER Standardized Curriculum was developed by NCCER in partnership with Pearson Education, Inc., the world's largest educational publisher.

Some features of NCCER's Standardized Curriculum are as follows:

- An industry-proven record of success
- Curricula developed by the industry for the industry
- National standardization providing portability of learned job skills and educational credits
- Compliance with the Office of Apprenticeship requirements for related classroom training (*CFR 29:29*)
- Well-illustrated, up-to-date, and practical information

NCCER also maintains a National Registry that provides transcripts, certificates, and wallet cards to individuals who have successfully completed a level of training within a craft in NCCER's Curriculum. *Training programs must be delivered by an NCCER Accredited Training Sponsor in order to receive these credentials.*

Special Features

In an effort to provide a comprehensive, user-friendly training resource, we have incorporated many different features for your use. Whether you are a visual or hands-on learner, this book will provide you with the proper tools to get started in the solar photovoltaic installation industry.

Introduction

This page is found at the beginning of each module and lists the Objectives, Performance Tasks, Trade Terms, and Prerequisites for that module. The Objectives list the skills and knowledge you will need in order to complete the module successfully. The Performance Tasks give you the opportunity to apply your knowledge to the real world duties that solar photovoltaic installers perform, including system design, irradiation measurement, installation, and maintenance. The list of Trade Terms identifies important terms you will need to know by the end of the module. The recommended Prerequisites suggests other titles from the Contren® Learning Series that will provide a helpful background of related information.

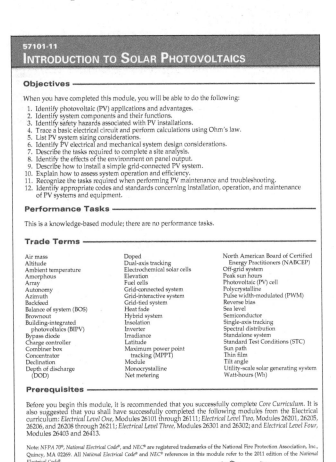

Color Illustrations and Photographs

Full-color illustrations and photographs are used throughout each module to provide vivid detail. These figures highlight important concepts from the text and provide clarity for complex instructions. Each figure reference is denoted in the text in *italic type* for easy reference.

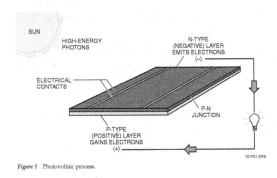

Figure 1 Photovoltaic process.

Notes, Cautions, and Warnings

Safety features are set off from the main text in highlighted boxes and are organized into three categories based on the potential danger of the issue being addressed. Notes simply provide additional information on the topic area. Cautions alert you of a danger that does not present potential injury but may cause damage to equipment. Warnings stress a potentially dangerous situation that may cause injury to you or a co-worker.

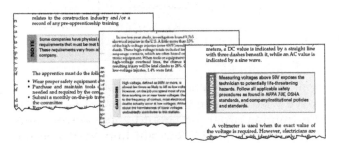

Did You Know?

The Did You Know? features offer hints, tips, and other helpful bits of information from the trade.

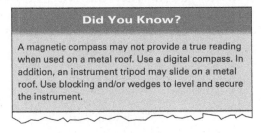

On Site

On Site features provide a head start for those entering the solar photovoltaic installation field by presenting technical tips and professional practices from solar PV installers in a variety of disciplines. On Site features often include real-life scenarios similar to those you might encounter on the job site.

On Site

Emergency Shelters

Emergency shelters are starting to incorporate PV systems to ensure power in the event of a disaster. Depending on the location, government rebates and/or grants may be available to offset the costs of building qualified structures.

Think About It

Think About It features use "What if?" questions to help you apply theory to real-world experiences and put your ideas into action.

Think About It

Fastener Checks

Some installers use permanent marker to draw a line over the center of the fastener when it is initially installed and after it has been checked for proper torque. That way, they can tell at a glance when fasteners have become loose. Can you find the loosened fastener in this picture?

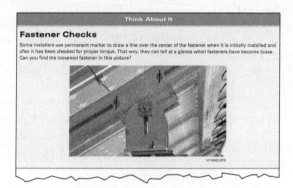

1010A06.EPS

Step-by-Step Instructions

Step-by-step instructions are used throughout to guide you through technical procedures and tasks from start to finish. These steps show you not only how to perform a task but also how to do it safely and efficiently.

phase, wye-connected motor. To identify the terminals of a wye-connected motor, proceed as follows:

Step 1 Taking the group of three common leads, arbitrarily identify them as leads 7, 8, and 9.

Step 2 Using the diagram in *Figure 64* as a guide, connect the positive lead from the battery to lead 7. Connect the lead from the battery negative through the N.O. switch to leads 8 and 9 simultaneously.

Step 3 Connect one of the three remaining lead pairs to the voltmeter terminals.

Step 4 Close the N.O. switch while observing the DC voltmeter. Use the lowest scale practical without over-ranging the meter. If

Trade Terms

Each module presents a list of Trade Terms that are discussed within the text and defined in the Glossary at the end of the module. These terms are denoted in the text with **blue bold type** upon their first occurrence. To make searches for key information easier, a comprehensive Glossary of Trade Terms from all modules is located at the back of this book.

Connections to the earth are made with grounding electrodes, ground grids, and ground mats. The resistance of these devices varies proportionately with the earth's resistivity, which in turn depends on the composition, compactness, temperature, and moisture content of the soil.

A good grounding system limits system-to-ground resistance to an acceptably low value. This protects personnel from a dangerously high voltage during a fault in the equipment. Furthermore, equipment damage can be limited by using this ground current to operate protective devices.

Ground testers measure the ground resistance of a grounding electrode or ground grid system. Some of the major purposes of ground testing are to verify the adequacy of a new grounding system, detect changes in an existing system, and determine the presence of hazardous step voltage and touch voltage.

Review Questions

Review Questions are provided to reinforce the knowledge you have gained. This makes them a useful tool for measuring what you have learned.

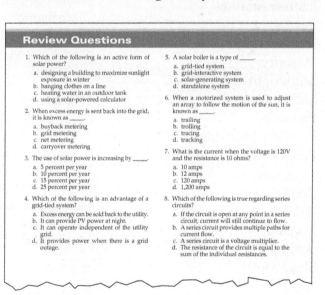

Review Questions

1. Which of the following is an active form of solar power?
   a. designing a building to maximize sunlight exposure in winter
   b. hanging clothes on a line
   c. heating water in an outdoor tank
   d. using a solar-powered calculator

2. When excess energy is sent back into the grid, it is known as _____.
   a. buyback metering
   b. grid metering
   c. net metering
   d. carryover metering

3. The use of solar power is increasing by _____.
   a. 5 percent per year
   b. 10 percent per year
   c. 15 percent per year
   d. 25 percent per year

4. Which of the following is an advantage of a grid-tied system?
   a. Excess energy can be sold back to the utility.
   b. It can provide PV power at night.
   c. It can operate independent of the utility grid.
   d. It provides power when there is a grid outage.

5. A solar boiler is a type of _____.
   a. grid-tied system
   b. grid-interactive system
   c. solar-generating system
   d. standalone system

6. When a motorized system is used to adjust an array to follow the motion of the sun, it is known as _____.
   a. trailing
   b. trolling
   c. tracing
   d. tracking

7. What is the current when the voltage is 120V and the resistance is 10 ohms?
   a. 10 amps
   b. 12 amps
   c. 120 amps
   d. 1,200 amps

8. Which of the following is true regarding series circuits?
   a. If the circuit is open at any point in a series circuit, current will still continue to flow.
   b. A series circuit provides multiple paths for current flow.
   c. A series circuit is a voltage multiplier.
   d. The resistance of the circuit is equal to the sum of the individual resistances.

NCCER Standardized Curricula

NCCER's training programs comprise more than 80 construction, maintenance, pipeline, and utility areas and include skills assessments, safety training, and management education.

Boilermaking
Cabinetmaking
Carpentry
Concrete Finishing
Construction Craft Laborer
Construction Technology
Core Curriculum:
   Introductory Craft Skills
Drywall
Electrical
Electronic Systems Technician
Heating, Ventilating, and
   Air Conditioning
Heavy Equipment Operations
Highway/Heavy Construction
Hydroblasting
Industrial Coating and Lining
   Application Specialist
Industrial Maintenance
   Electrical and
   Instrumentation Technician
Industrial Maintenance
   Mechanic
Instrumentation
Insulating
Ironworking
Masonry
Millwright
Mobile Crane Operations
Painting
Painting, Industrial
Pipefitting
Pipelayer
Plumbing
Reinforcing Ironwork
Rigging
Scaffolding
Sheet Metal
Signal Person
Site Layout
Sprinkler Fitting
Tower Crane Operator
Welding

Green/Sustainable Construction

Building Auditor
Fundamentals of
   Weatherization
Introduction to Weatherization
Sustainable Construction
   Supervisor
Weatherization Crew Chief
Weatherization Technician
Your Role in the Green
   Environment

Energy

Introduction to the Power
   Industry
Introduction to Solar
   Photovoltaics
Introduction to Wind Energy
Power Industry Fundamentals
Power Generation Maintenance
   Electrician
Power Generation I&C
   Maintenance Technician
Power Generation Maintenance
   Mechanic
Power Line Worker
Solar Photovoltaic Systems
   Installer

Pipeline

Control Center Operations,
   Liquid
Corrosion Control
Electrical and Instrumentation
Field Operations, Liquid
Field Operations, Gas
Maintenance
Mechanical

Safety

Field Safety
Safety Orientation
Safety Technology

Management

Fundamentals of Crew
   Leadership
Project Management
Project Supervision

Supplemental Titles

Applied Construction Math
Careers in Construction
Tools for Success

Spanish Translations

Rigging Fundamentals
   (Principios Básicos de
   Maniobras)
Carpentry Fundamentals
   (Introducción a la
   Carpintería, Nivel Uno)
Carpentry Forms
   (Formas para Carpintería,
   Nivel Trés)
Concrete Finishing, Level One
   (Acabado de Concreto,
   Nivel Uno)
Core Curriculum:
   Introductory Craft Skills
   (Currículo Básico:
   Habilidades Introductorias
   del Oficio)
Drywall, Level One
   (Paneles de Yeso, Nivel Uno)
Electrical, Level One
   (Electricidad, Nivel Uno)
Field Safety
   (Seguridad de Campo)
Insulating, Level One
   (Aislamiento, Nivel Uno)
Masonry, Level One
   (Albañilería, Nivel Uno)
Pipefitting, Level One
   (Instalación de Tubería
   Industrial, Nivel Uno)
Reinforcing Ironwork, Level One
   (Herreria de Refuerzo,
   Nivel Uno)
Safety Orientation
   (Orientación de Seguridad)
Scaffolding
   (Andamios)
Sprinkler Fitting, Level One
   (Instalación de Rociadores,
   Nivel Uno)

Acknowledgments

This curriculum was revised as a result of the farsightedness and leadership of the following sponsors:

NeoSolvis Engineering C.S.P
Pumba Electric LLC
Westside Technical Center
Solar Source Institute

Tri-City Electrical Contractors, Inc.
Marion Technical Institute
Industrial Management & Training Institute, Inc.
Florida Solar Energy Center

This curriculum would not exist were it not for the dedication and unselfish energy of those volunteers who served on the Authoring Team. A sincere thanks is extended to the following:

Nicolás Estévez
L. J. LeBlanc
Joseph S. Lowe
Mark Maher

Mike Powers
Antonio Vazquez
Marcel Veronneau

A final note: This book is the result of a collaborative effort involving the production, editorial, and development staff at Pearson Education, Inc., and the NCCER. Thanks to all of the dedicated people involved in the many stages of this project.

NCCER Partners

American Fire Sprinkler Association
Associated Builders and Contractors, Inc.
Associated General Contractors of America
Association for Career and Technical Education
Association for Skilled and Technical Sciences
Carolinas AGC, Inc.
Carolinas Electrical Contractors Association
Center for the Improvement of Construction
    Management and Processes
Construction Industry Institute
Construction Users Roundtable
Design Build Institute of America
Merit Contractors Association of Canada
Metal Building Manufacturers Association
NACE International
National Association of Manufacturers
National Association of Minority Contractors
National Association of Women in Construction
National Insulation Association
National Ready Mixed Concrete Association
National Technical Honor Society
National Utility Contractors Association
NAWIC Education Foundation

North American Technician Excellence
Painting & Decorating Contractors of America
Portland Cement Association
Skills USA
Steel Erectors Association of America
U.S. Army Corps of Engineers
University of Florida, M.E. Rinker School of
    Building Construction
Women Construction Owners & Executives, USA

Contents

Note: *NFPA 70®*, *National Electrical Code®*, and *NEC®* are registered trademarks of the National Fire Protection Association, Quincy, MA 02269. All *National Electrical Code®* and *NEC®* references in this textbook refer to the 2011 edition of the *National Electrical Code®*.

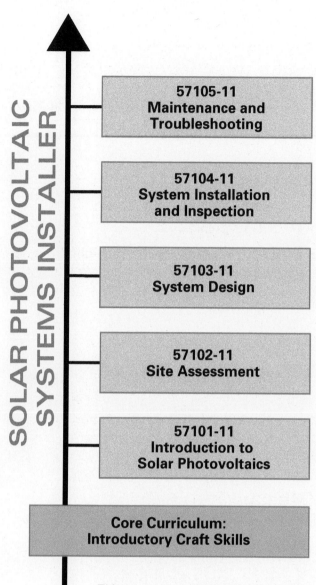

SOLAR PHOTOVOLTAIC SYSTEMS INSTALLER

57105-11
Maintenance and Troubleshooting

57104-11
System Installation and Inspection

57103-11
System Design

57102-11
Site Assessment

57101-11
Introduction to Solar Photovoltaics

Core Curriculum:
Introductory Craft Skills

This course map shows all of the modules in *Solar Photovoltaic Systems Installer*. The suggested training order begins at the bottom and proceeds up. Skill levels increase as you advance on the course map. The local Training Program Sponsor may adjust the training order.

Introduction to Solar Photovoltaics

57101-11

Trainees with successful module completions may be eligible for credentialing through NCCER's National Registry. To learn more, go to **www.nccer.org** or contact us at **1.888.622.3720.** Our website has information on the latest product releases and training, as well as online versions of our Cornerstone magazine and Pearson's product catalog.

Your feedback is welcome. You may email your comments to **curriculum@nccer.org,** send general comments and inquiries to **info@nccer.org,** or use the User Update form at the back of this module.

V.2 11/13

Objectives

When you have completed this module, you will be able to do the following:

1. Identify photovoltaic (PV) applications and advantages.
2. Identify system components and their functions.
3. Identify safety hazards associated with PV installations.
4. Trace a basic electrical circuit and perform calculations using Ohm's law.
5. List PV system sizing considerations.
6. Identify PV electrical and mechanical system design considerations.
7. Describe the tasks required to complete a site analysis.
8. Identify the effects of the environment on panel output.
9. Describe how to install a simple grid-connected PV system.
10. Explain how to assess system operation and efficiency.
11. Recognize the tasks required when performing PV maintenance and troubleshooting.
12. Identify appropriate codes and standards concerning installation, operation, and maintenance of PV systems and equipment.

Performance Tasks

This is a knowledge-based module; there are no performance tasks.

Trade Terms

Air mass
Altitude
Ambient temperature
Amorphous
Array
Autonomy
Azimuth
Backfeed
Balance of system (BOS)
Brownout
Building-integrated photovoltaics (BIPV)
Bypass diode
Charge controller
Combiner box
Concentrator
Declination
Depth of discharge (DOD)

Doped
Dual-axis tracking
Electrochemical solar cells
Elevation
Fuel cells
Grid-connected system
Grid-interactive system
Grid-tied system
Heat fade
Hybrid system
Insolation
Inverter
Irradiance
Latitude
Maximum power point tracking (MPPT)
Module
Monocrystalline
Net metering

North American Board of Certified Energy Practitioners (NABCEP)
Off-grid system
Peak sun hours
Photovoltaic (PV) cell
Polycrystalline
Pulse width-modulated (PWM)
Reverse bias
Sea level
Semiconductor
Single-axis tracking
Spectral distribution
Standalone system
Standard Test Conditions (STC)
Sun path
Thin film
Tilt angle
Utility-scale solar generating system
Watt-hours (Wh)

Prerequisites

Before you begin this module, it is recommended that you successfully complete *Core Curriculum*. It is also suggested that you shall have successfully completed the following modules from the Electrical curriculum: *Electrical Level One*, Modules 26101 through 26111; *Electrical Level Two*, Modules 26201, 26205, 26206, and 26208 through 26211; *Electrical Level Three*, Modules 26301 and 26302; and *Electrical Level Four*, Modules 26403 and 26413.

Note: *NFPA 70®*, *National Electrical Code®*, and *NEC®* are registered trademarks of the National Fire Protection Association, Inc., Quincy, MA 02269. All *National Electrical Code®* and *NEC®* references in this module refer to the 2011 edition of the *National Electrical Code®*.

Contents

Topics to be presented in this module include:

Figures and Tables

Figures and Tables (*continued*) ───────

1.0.0 INTRODUCTION

Solar power has been in use for over two thousand years. The ancient Greeks used thick walls to trap heat during the day and release it slowly at night. They also oriented buildings to provide shading in summer while maximizing sunlight in winter. Solar energy was also used to preserve food, heat water, and dry clothes. These early applications of solar power were very effective and are still in use today. They are known as passive forms of solar power because they use sunlight without transforming it.

In 1767, Horace de Saussure experimented with glass boxes to determine how much heat could be trapped by the glass. He discovered that heat could be collected on sunny days even if the outdoor temperature was quite low, as on a mountaintop. These experiments helped scientists understand the effects of atmospheric differences on outdoor temperature and also became the basis for passive solar collectors and solar ovens.

Active forms of solar power did not come into use until 1839, when Alexander Becquerel discovered that certain materials produce electric current when exposed to light. This eventually led to the invention of the first photovoltaic (PV) cell. In 1891, Clarence Kemp patented the first water heater powered by solar energy. The first practical PV cell was invented by Bell Laboratories in 1954 using modified silicon cells.

A PV cell consists of two layers of semiconductor, one p-type (positive) and the other n-type (negative). The electrical contacts are screened in a grid on both sides of the panel. High-energy light

particles known as photons strike the semiconductor atoms and release free electrons, producing current (*Figure 1*). Many cells are combined in a solar panel, and these modular panels are connected in an array.

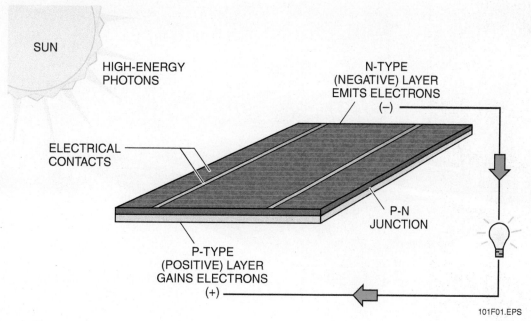

SUN

HIGH-ENERGY PHOTONS

N-TYPE (NEGATIVE) LAYER EMITS ELECTRONS (–)

ELECTRICAL CONTACTS

P-N JUNCTION

P-TYPE (POSITIVE) LAYER GAINS ELECTRONS (+)

101F01.EPS

Figure 1   Photovoltaic process.

As America entered the space race in the 1950s, scientists realized that solar arrays provided an ideal power source for satellites. Later, solar cells became popular for operating rural telephones and radio transmitters, and for homes that were too remote to be served by the electrical grid.

In the 1970s, the energy crisis focused more attention on alternative energy sources, and the use of solar power increased in both residential and commercial applications (*Figure 2*). Today, solar power is used for everything from hand-held calculators to giant solar farms that produce enough energy to operate a small city.

PV power has many advantages over other energy sources. Sunlight is a limitless resource, and most areas can support some degree of solar power. PV power is clean and environmentally friendly. It can be harnessed without disrupting the environment and produces no hazardous waste or emissions. It is also quiet, reliable, and requires little maintenance. In addition, PV power can be generated onsite, and does not have the mining/drilling or transportation requirements of fossil fuels. The power generated is also a domestic product, which strengthens the economy, produces jobs, and reduces reliance on foreign energy.

PV systems help to offset power use during peak demand periods. The electrical grid typically reaches peak demand during the late afternoon when people come home from work and turn on the air conditioning. For a solar-powered facility, this is a time when it is producing energy. Putting energy back into the grid during peak demand periods reduces the possibility of brownouts. As electricity is produced, it is used at the home or business first, and then any excess electricity is sent into the grid. This is known as net metering. Customers receive credits for generating power, which can help to offset the initial cost of installing a PV system. In addition, customers often receive government incentives and utility rebates when installing these systems.

The use of solar power is now increasing by more than 25 percent per year, creating a huge demand for certified installers. This module covers the basic concepts of PV systems and their components. It also explains how PV systems are sized, designed, and installed. Successful completion of this module will help prepare trainees for the North American Board of Certified Energy Practitioners (NABCEP) PV Entry Level Exam. Trainees who pass the NABCEP exam may work as helpers installing PV systems. After gaining additional training and logging the required number of installations, trainees will then be qualified to take the NABCEP Certified PV Installer Exam.

101F02.EPS

Figure 2 Typical commercial PV system.

2.0.0 APPLICATIONS

PV arrays can be installed using a wide variety of mounting methods. Some of these include roof mounting, pole mounting, ground mounting, and wall mounting. They can also be installed as part of a shade structure, such as a carport. In addition to mounting methods, PV arrays are also classified by how they are connected to other power sources and loads. PV systems can operate connected to or independent of the utility grid. They can also be connected to other energy sources, such as wind turbines, and energy storage systems, such as batteries. Solar energy can be classified into four basic types of systems: standalone systems, grid-connected systems, grid-interactive systems, and utility-scale solar generating systems.

2.1.0 Standalone Systems

Standalone PV systems can be either direct-drive or battery-powered. These systems are commonly used to provide power in areas where access to the grid is inconvenient or unavailable.

Direct-drive systems power DC loads, such as ventilation fans, irrigation pumps, and remote cattle watering systems. Because these systems operate only when the sun is shining, it is essential to match the power output of the array to the load.

Battery-powered standalone systems use PV energy to charge one or more batteries. The battery system then supplies either a DC load or an AC load when an inverter is used to convert the DC to AC. Battery-powered standalone systems range from handheld electronics to trailer-mounted systems used to provide emergency power (*Figure 3*). They can also be used for remote data monitoring, emergency highway signage, and street or parking lot lighting (*Figure 4*).

Standalone systems are also used to power buildings in remote areas. These systems do not require power poles and transmission lines, and therefore eliminate the possibility of transmission failures due to downed power lines. These are known as off-grid systems. Off-grid systems use batteries for energy storage as well as battery-based inverter systems. Charge controllers are used to maximize the battery-charging efficiency of the solar array. Backup power is provided by an engine-driven generator.

On Site

Zero Energy Homes

Some homes use a combination of active and passive solar energy, along with weatherization and other techniques, to produce as much energy as they use. They are known as zero energy homes, one of which is shown here. This home is in Denver, Colorado. The roof overhangs are designed to shade the windows from the sun in the summer, but allow sunlight to provide light and warmth in the winter when the angle of the sun is lower. Over the course of a year, the energy produced by the solar array equals the energy required to operate the home, resulting in a net use of zero.

101SA02.EPS

Did You Know?

The Earth receives more energy from the sun in a single hour than the world uses in a whole year.

Source: Massachusetts Institute of Technology

101F03.EPS

Figure 3   Portable PV generator.

2.2.0 Grid-Connected Systems

Grid-connected systems operate in parallel with the utility grid. Also known as grid-tied systems, these systems are designed to provide supplemental power to the building or residence (*Figure 5*). Since they are tied to the utility, they only operate when grid power is available.

Grid-tied systems invert the DC produced by a solar array into AC, which is then sent to the building's electrical panel to supply power. A DC disconnect is required between the array and the inverter, while an AC disconnect is required between the inverter and the building service panel.

Grid-tied systems require the installation of a special meter by the utility. During the daytime, any power in excess of the load is sold back to the utility in the form of credits. This can be shown by the building's electrical meter, which effectively runs backward when excess energy is supplied. At night and during periods when the load exceeds the system output, the required power is supplied by the electric utility.

101F04.EPS

Figure 4 Solar-powered parking lot fixture.

On Site

Emergency Shelters

Emergency shelters are starting to incorporate PV systems to ensure power in the event of a disaster. Depending on the location, government rebates and/or grants may be available to offset the costs of building qualified structures.

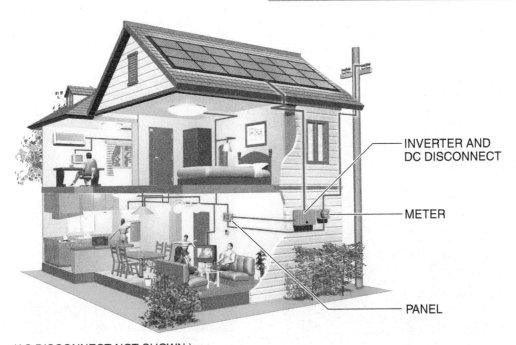

INVERTER AND
DC DISCONNECT

METER

PANEL

(AC DISCONNECT NOT SHOWN.)

101F05.EPS

Figure 5 Grid-tied PV system.

NCCER — *Contren® Learning Series* 57101-11

The inverter allows for the conversion of power and also shuts down the system during a power outage or other electrical failure. This is a safety feature that prevents voltage from traveling back into the grid via backfeed.

2.3.0 Grid-Interactive Systems

Like a grid-tied system, a grid-interactive system is connected to the utility and uses inverted PV power as a supplement. Unlike a grid-tied system, which is literally tied to the grid and cannot function independently, a grid-interactive system provides a means of independent power. Grid-interactive systems include batteries that can supply power during outages and after sundown.

When the PV system operates, it first charges the batteries, then it satisfies the existing load, and then any excess power is sent to the grid. As with off-grid systems, a charge controller is used to monitor the battery charge to ensure consistent power with minimal downtime. In the event of a power outage, battery power is supplied to critical loads through the use of a designated subpanel. When these systems are used with other energy sources, such as wind turbines or generators, they are known as hybrid systems.

2.4.0 Utility-Scale Solar Generating Systems

There are two types of utility-scale solar generating systems. The first type is a steam generator that heats the water using concentrated sunlight rather than fossil fuel. This type of generating system is known as a solar boiler (*Figure 6*). The second type uses a large bank of solar cells to produce direct current (*Figure 7*). Motorized systems are often used to adjust the panel position so that it follows the movement of the sun throughout the day. This is known as tracking.

Utility-scale solar generating systems are not yet in common use, but will become increasingly popular as the technology advances. This module focuses on grid-tied and grid-interactive systems used in residential and commercial applications.

3.0.0 OHM'S LAW AND POWER

An understanding of Ohm's law is essential when determining the power and load requirements of PV systems. Ohm's law defines the relationship between current (I), voltage or electromotive force (E), and resistance (R). There are three ways to express Ohm's law mathematically.

- The current in a circuit is equal to the applied voltage divided by the resistance:

$$I = \frac{E}{R}$$

- The resistance of a circuit is equal to the applied voltage divided by the current:

$$R = \frac{E}{I}$$

Figure 6   Sun-tracking solar boiler.

101F06.EPS

 Introduction to Solar Photovoltaics

Figure 7   Solar power plant.

The voltage applied to a circuit is equal to the product of the current and the resistance:

$$E = I \times R$$

If any two of the quantities are known, the third can be calculated. The Ohm's law equations can be practiced using an Ohm's law circle, as shown in *Figure 8*. To find the equation for E, I, or R when two quantities are known, cover the unknown third quantity. The other two quantities in the circle will indicate how the covered quantity may be found.

Example 1:

Find I when E = 120V and R = 30 ohms.

$$I = \frac{E}{R}$$

$$I = \frac{120V}{30 \text{ ohms}}$$

$$I = 4A$$

Example 2:

Find R when E = 240V and I = 20A.

$$R = \frac{E}{I}$$

$$R = \frac{240V}{20A}$$

$$R = 12 \text{ ohms}$$

Example 3:

Find E when I = 15A and R = 8 ohms.

$$E = I \times R$$

$$E = 15A \times 8 \text{ ohms}$$

$$E = 120V$$

$$I = \frac{E}{R} \qquad E = I \times R \qquad R = \frac{E}{I}$$

	LETTER SYMBOL	UNIT OF MEASUREMENT
CURRENT	I	AMPERES (A)
RESISTANCE	R	OHMS (Ω)
VOLTAGE	E	VOLTS (V)

101F08.EPS

Figure 8   Ohm's law circle.

NCCER — *Contren® Learning Series*   57101-11

3.1.0 Applying Ohm's Law to Series and Parallel Circuits

You will recall that loads can be arranged in series, in parallel, or in series-parallel. *Figure 9* shows these three types of circuits.

3.1.1 Series Circuits

A series circuit provides only one path for current flow and is a voltage divider. The total resistance of the circuit is equal to the sum of the individual resistances. The 12V series circuit in *Figure 9* has two 30-ohm loads. The total resistance is therefore 60 ohms. The amount of current flowing in the circuit is calculated as follows:

$$I = \frac{E}{R}$$

$$I = \frac{12V}{60 \text{ ohms}}$$

$$I = 0.2A$$

The current flow is the same through all the loads. The voltage measured across any load (voltage drop) depends on the resistance of that load. The sum of the voltage drops equals the total voltage applied to the circuit. An important trait of a series circuit is that if the circuit is open at any point, no current will flow.

3.1.2 Parallel Circuits

In a parallel circuit, each load is connected directly to the voltage source; therefore, the voltage drop is the same through all loads and current is divided between the loads. The source sees the circuit as two or more individual circuits containing one load each. In the parallel circuit in *Figure 9*, the source sees three circuits, each containing a 30-ohm load. The current flow through any load is determined by the resistance of that load. Therefore, the total current drawn by the circuit is the sum of the individual currents. The total resistance of a parallel circuit is calculated differently from that of a series circuit. In a parallel circuit, the total resistance is less than the smallest of the individual resistances.

For example, each of the 30-ohm loads draws 0.4A at 12V; therefore, the total current is 1.2A:

$$I = \frac{E}{R}$$

$$I = \frac{120V}{30 \text{ ohms}}$$

$$I = 0.4A \times 3 = 1.2A$$

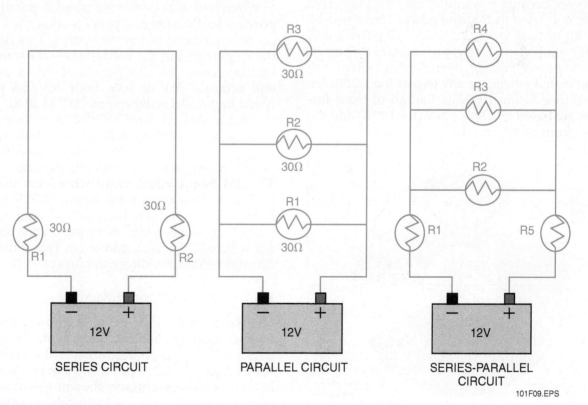

Figure 9 Types of circuits.

101F09.EPS

Now, Ohm's law can be used again to calculate the total resistance:

$$R = \frac{E}{I}$$

$$R = \frac{12V}{1.2A}$$

$$R = 10 \text{ ohms}$$

3.1.3 Series-Parallel Circuits

When loads are connected in series-parallel, the parallel loads must first be converted to their equivalent resistances. Then the load resistances are added to determine the total circuit resistance.

3.2.0 Ohm's Law and Power

Power is the ability to do work. Mechanical power is often expressed in horsepower (hp). In electrical circuits, power is measured in watts (W). One hp equals 746 watts. One watt is the power used when one ampere of current flows through a potential difference of one volt. For this reason, watts are referred to as volt-amps (VA). Power (P) is determined by multiplying the rated current (I) by the rated voltage (E):

$$P = I \times E$$

This equation is sometimes called Ohm's law for power, because it is similar to Ohm's law. This equation is used to find the power consumed by a circuit or load when the values of current and voltage are known. Using variations of this equation, the power, voltage, or current in a circuit can be calculated whenever any two of the values are known. See *Figure 10*. Note that all of these formulas are based on Ohm's law (E = I × R) and the power formula (P = I × E).

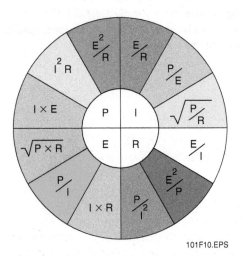

101F10.EPS

Figure 10 Expanded Ohm's law circle.

Watt-hours (Wh) are calculated by multiplying the power in watts (VA) by the number of hours during which the power is used. The kilowatt-hour (kWh) is commonly used for larger amounts of electrical work or energy. (The prefix *kilo* means one thousand.) For example, if a light bulb uses 100W or 0.1kW for 10 hours, the amount of energy consumed is 0.1kW × 10 hours = 1.0kWh.

Very large amounts of electrical work or energy are measured in megawatts (MW). (The prefix *mega* means one million.)

3.3.0 Series and Parallel Circuits in Solar PV Systems

Household circuits and most other loads are wired in parallel. However, solar panels can be wired in series, in parallel, or in series-parallel, depending on the desired output. Wiring solar panels in series increases the voltage, while wiring them in parallel increases the amperage.

When solar panels are wired in series, the positive terminal of one panel is connected to the negative terminal of another. Unlike loads that drop voltage and use power, when you put solar panels in series they are creating power. This doubles the voltage, but the total amperage remains the same. For example, a 36V/5A panel delivers 180W of power (VA). Two 36V/5A panels wired in series will produce 72V at 5A or 360W. Four 36V/5A panels wired in series will produce 144V at 5A or 720W.

When two solar panels are wired in parallel, the positive terminal of one panel is connected to the positive terminal of the next panel. This doubles the amperage, but the voltage stays the same. For example, two 36V/5A panels wired in parallel will produce 36V at 10A. Four 36V/5A panels wired in parallel will produce 36V at 20A.

Solar arrays can be connected in series-parallel to achieve the desired voltage and current of the total array (*Figure 11*). For example, two 36V/5A panels can be wired in series to produce 72V at 5A, and then connected to two panels wired in parallel to produce a total output of 72V at 10A. If these four panels are then parallel-connected to four other panels wired in the same way, the voltage will be 72V at 20A, and so on. This is the basic concept behind building solar arrays.

3.4.0 Peak Sun and Power

To determine the total panel or array output, the wattage is multiplied by the peak sun hours per day for the geographical area to determine the watt-hours produced per day. Peak sun hours or insolation values represent the equivalent number of hours per day when solar irradiance averages 1,000W/m². Irradiance is a measure of radiation

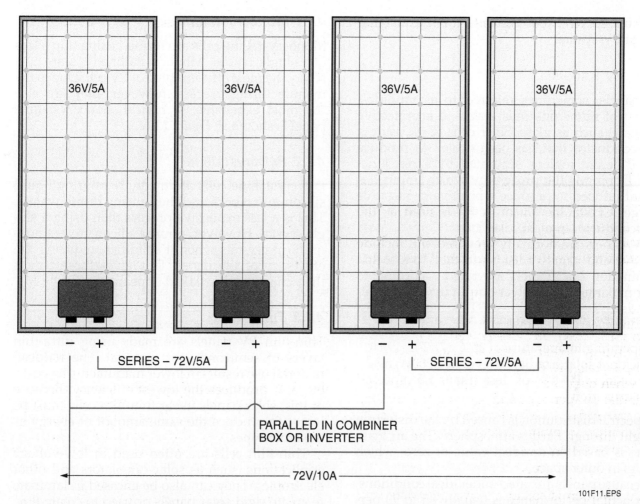

Figure 11 Series-parallel wiring in a solar array.

density and varies widely by location. Areas with few cloudy days and low levels of dust have high levels of irradiance. In addition, higher elevations have greater irradiance than those at sea level. Higher elevations also have cooler temperatures, which reduces the resistance and increases the array output voltage.

The peak sun hours for a given location represent an average value since solar intensity varies by time of day, season, and cloud cover. For example, a location may receive 800W/m² for three hours and 1,200W/m² for two hours. The total would be 4,800W/m² divided by 1,000W/m² = 4.8 peak sun hours.

To convert peak sun hours to power, multiply the wattage by the peak sun hours. For example, the power produced by a 36V/5A solar panel is 36V × 5A = 180W. If this 180W panel receives five peak sun hours per day, it will produce 180W × 5 = 900 watt-hours (0.9kWh) per day.

> **NOTE**
>
> Insolation maps and other resources can be found on the National Renewable Energy Laboratory (NREL) website at www.nrel.gov.

Solar Energy

The amount of solar radiation that reaches the Earth's outer atmosphere is nearly constant at 1,360W/m². On a clear day, approximately 70 percent of this radiation travels through the atmosphere to the Earth. Average values are typically 1,000W/m² or lower, but can be as high as 1,500W/m² when magnified by certain atmospheric conditions.

4.0.0 PV SYSTEM COMPONENTS

A typical grid-interactive PV system has four main components: the panels, an inverter, batteries, and a charge controller. The remaining components are known as the balance-of-system (BOS) components. They include the panel mounts, wiring, overcurrent protection, grounding system, and disconnects. PV installations must comply with all applicable requirements of the *National Electrical Code®* (NEC®). *NEC Article 690*

contains specific requirements for the installation of PV systems.

4.1.0 PV Panels

PV panels, sometimes referred to as modules, consist of numerous cells sealed in a protective laminate, such as glass. Solar cells work using a semiconductor that has been doped to produce two different regions separated by a p-n junction. Doping is the process by which impurities are introduced to produce a positive or negative charge. Crystalline silicon (c-Si) is used as the semiconductor in most solar cells.

PV panels are normally prewired and include positive and negative leads attached to a sealed termination box. Panels are rated according to their maximum DC power output in watts under Standard Test Conditions (STC). Standard Test Conditions are as follows:

- Operating temperature of 25°C (77°F)
- Incident solar irradiance level of 1,000W/m²
- Air mass (AM) of less than 1.5 spectral distribution

Spectral distribution is caused by the distortion of light through Earth's atmosphere. The air mass value is based on an ideal value of zero, which occurs in outer space.

Because panels are rated under ideal conditions, their actual performance is usually 85 to 90 percent of the STC rating. The output of a PV cell depends on its efficiency, cleanliness, orientation, amount of sunlight, and temperature. Temperature increases can cause a significant decrease in the system output voltage. This is because higher air temperatures decrease irradiance. The irradiance drops by 50 percent for every 10°C temperature rise (18°F). (Note that this isn't a straight conversion because it represents an interval, rather than a defined temperature.)

PV cells are characterized by the type of crystal used in them. There are three basic types of PV cells: monocrystalline, polycrystalline, and amorphous (commonly known as thin film). The type of crystal used determines the efficiency of the cell. For example, a typical monocrystalline panel might produce 75W of power, while an equivalent polycrystalline cell might produce 65W, and a thin-film cell might produce 45W.

> **NOTE**
> Actual panel wattages vary widely depending on manufacturer, panel size, and type of semiconductor. Always consult the nameplate data for the specific panel in use.

4.1.1 Monocrystalline

Monocrystalline cells are formed using thin slices of a single crystal. Monocrystalline cells are currently the most efficient type of PV cell. Due to the manufacturing process, however, they are also the most expensive. A typical monocrystalline panel is shown in *Figure 12*.

4.1.2 Polycrystalline

Polycrystalline cells are made by pouring liquid silicon into blocks and then slicing it into wafers. This is a less expensive process than using a single crystal. However, poured silicon creates non-uniform crystals when it solidifies, reducing the efficiency of the panel. This gives these panels their characteristic flaked appearance (*Figure 13*).

4.1.3 Thin Film

Thin-film PV panels are made using ultra-thin layers of semiconductor material. The reduced material use results in lower manufacturing costs, but also produces the lowest efficiency. Because of this, solar panels using thin-film cells must be larger to produce the same amount of energy as the other types.

Thin-film cells are often used in low-voltage applications, such as solar calculators and other electronics. They can also be encased in laminate to create rigid solar panels or used to create flexible building materials, such as for alternate rows

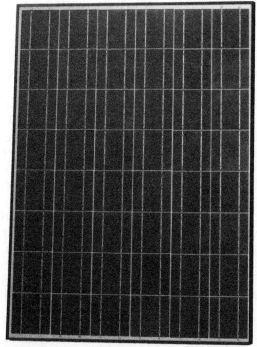

101F13.EPS

Figure 12 Monocrystalline PV panel.

Figure 13 Polycrystalline PV panel.

101F13.EPS

in roofing shingles (*Figure 14*). When a solar panel is built into a structure, it is known as building-integrated photovoltaics (BIPV).

4.2.0 Inverters

Inverters are used to convert the DC produced by the PV array into AC that can be used by various loads (*Figure 15*). As you will recall, AC travels in a sine wave. There are two common types of inverters: modified sine wave and true sine wave. Modified sine wave inverters are less expensive but do not provide the power quality of true sine wave inverters. They are not recommended for use with electronic equipment or other sensitive devices, including certain types of motors. True sine wave inverters produce smooth power similar to that provided by the grid. The quality of this power may actually be superior to grid power as it is free of voltage dips, spikes, and noise.

Inverters are available in a wide variety of types and load ratings. Some inverters are designed to operate with standalone systems, while others are designed for grid-tied and grid-interactive systems. Off-grid systems use battery-based inverter systems with an output voltage of 120/240VAC so they can operate larger loads such as water heaters and stoves. Inverters are rated in both continuous watts and surge watts. Some inverters include integral ground-fault protection and DC disconnects, while others require that these devices be installed separately. Many inverters use a digital display to indicate ambient temperature, voltage output, and other system parameters. The display is normally in standby mode and appears blank. To access various system outputs, simply tap on the inverter housing and the values are displayed.

4.3.0 Batteries

Batteries are used to store energy produced by the PV array and supply it to electrical loads as needed (*Figure 16*). Batteries are also used to operate the PV array near its maximum power point, to stabilize output voltages during periods of low production, and to supply startup currents to loads such as motors. All stationary battery installations must comply with the provisions of *NEC Article 480*.

THIN-FILM SOLAR PANEL

FLEXIBLE THIN FILM FOR BIPV APPLICATIONS

101F14.EPS

Figure 14   Typical thin-film applications.

101F15.EPS

Figure 15   True sine wave inverter.

101F16.EPS

Figure 16   12V PV Batteries.

Solar batteries require proper charging. While some deep cycle batteries can be discharged to 20 percent of capacity, it is best not to go that low as repeated deep cycling shortens battery life. The amount of charge remaining is known as the depth of discharge (DOD). Battery capacity is measured in amp-hours (Ah). For example, a 100Ah battery from which 80Ah has been withdrawn has undergone an 80 percent DOD. Some PV systems include a low battery warning light or cutoff switch to prevent the damage caused by repeated deep cycling.

Batteries must be protected from extreme temperatures. Cold temperatures reduce battery output while hot temperatures increase deterioration. With proper care and appropriate cycling, solar batteries can last between 5 and 7 years. Batteries are rated by the expected number of charge cycles as well as the maximum discharge current. Each cell produces 2V, so a 6V battery contains three cells. Batteries can be connected in series or in parallel to produce the desired output. Residential PV systems are limited to no more than 24 two-volt cells connected in series (48V total) per *NEC Section 690.71(B)(1)*.

> **NOTE**
>
> There are stringent code requirements for battery banks in excess of 48V. See *NEC Section 690.71* for details.

There are two types of solar batteries: flooded lead acid (FLA) and sealed absorbent glass mat (AGM).

4.3.1 FLA Batteries

FLA batteries are more common and less expensive than AGM batteries. However, they require regular maintenance to add water and prevent corrosion. FLA batteries can deteriorate quickly if the fluid levels drop and the plates oxidize. They require good air circulation to prevent gases from accumulating to explosive levels or causing corrosion of other equipment. They are normally located in a separate compartment with appropriate venting and warning labels.

> **WARNING!**
>
> Overcharging and improper venting can cause an explosive gas buildup within a battery. When this occurs, the battery may leak, feel warm, or exhibit signs of swelling. The pressure rises within the cells until a nearby spark or short circuit ignites the gas. Battery explosions release violent sprays of caustic acid and shrapnel that can cause severe injury or even death.

4.3.2 AGM Batteries

AGM batteries are sealed lead-acid batteries. The acid is contained in a special glass fiber mat so a vented battery compartment is not required. In addition, AGM batteries generally do not leak or produce corrosive gases. These batteries are also maintenance-free.

4.4.0 Charge Controllers

Charge controllers are used to regulate the charge and discharge of the system batteries. They are rated by their maximum AC current output, so it essential that they be sized to match the application. There are two main types of charge controllers: pulse width-modulated (PWM) and maximum power point tracking (MPPT). Older charge controllers use shunts or relay transistors and do not work well with sealed batteries. PWM charge controllers are ideal for use with sealed batteries but can only be used within a limited range of panel configurations and voltages. An MPPT charge controller is shown in *Figure 17*. These controllers harvest energy more efficiently and can be used with a wider range of array configurations and voltages. A digital display is provided to monitor power use, battery charge, and other system performance data. MPPT charge controllers are preferred for use in cold climates because they prevent the overvoltage that can occur at lower temperatures and provide more precise control over the DOD.

4.5.0 BOS Components

BOS components include the wiring, grounding system, disconnects, foundations, and support frames. They also include any required subpanels,

101F17.EPS

Figure 17   MPPT charge controller.

conduit, and combiner boxes. Weatherproof combiner boxes are used to connect strings of solar panels to create a larger array, and to provide a convenient array disconnect point.

4.5.1 Electrical System Components

A PV electrical system includes all wiring, devices, and components between the service entrance and the final panel termination. The most important safety devices in a PV system are the AC and DC disconnects and the grounding system.

Each piece of equipment in a PV system requires a switch or breaker to disconnect it from all sources of power per *NEC Section 690.15*. Disconnect requirements are listed in *NEC Section 690.17*. A DC disconnect is shown in *Figure 18*. This disconnect is attached to the inverter. If the inverter does not include a DC disconnect, it must be purchased and installed separately.

The AC disconnect is located near the service panel. A typical AC disconnect is shown in *Figure 19*. PV disconnects may be located indoors if the panel is in the building.

> **NOTE**
> PV disconnects are not allowed in residential bathrooms per *NEC Section 690.14(C)(1)*.

101F19.EPS

Figure 19   AC disconnect.

The grounding system is essential to the electrical integrity of a PV array. All exposed non–current-carrying metal parts of the equipment must be bonded and connected to ground. The ground connection must be continuous. The roof ground should be a minimum 6 AWG solid copper to provide extra strength. Always use approved cleats when making ground connections. They are designed to pierce the finish of the panel to ensure electrical continuity. A ground rod connection is shown in *Figure 20*.

Many ground-mounted PV systems incorporate multiple levels of protection against stray voltages, such as lightning. These are installed in addition to the equipment grounding system and may include surge protectors and the use of grounded metal fencing for ground-mounted arrays.

Ground-fault protection is required per *NEC Section 690.5*. Some inverters include integral ground-fault protection. If it is not supplied in the inverter, it must be located elsewhere in the PV system.

Other electrical system BOS components include the panel wiring, conduit, combiner boxes, and termination boxes. See *Figure 21*. All wiring must be sized for the load and the distance between components. All cables carrying other-than-low voltages must be routed in conduit.

4.5.2 Footers and Support Structures

Support frames and footers are the most labor-intensive portion of any ground-mounted installation. In a system installed in an open field, the

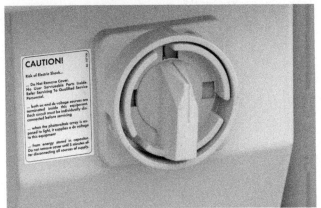

101F18.EPS

Figure 18   DC disconnect.

Figure 20 Grounding system.

101F20.EPS

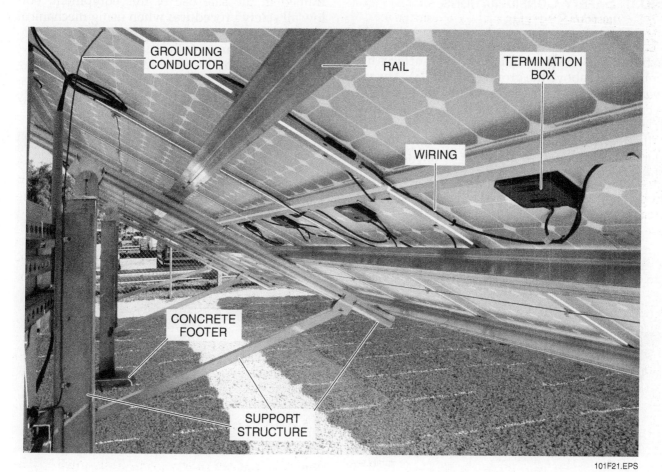

Figure 21 BOS components.

101F21.EPS

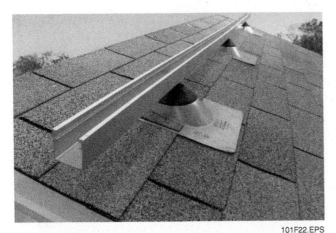

Figure 22   Roof-mounted rail support.

footers are typically concrete while the support frames are aluminum or steel.

Roof-mounted systems are mounted on aluminum or stainless steel rails secured into the roof trusses using lag bolts. See *Figure 22*. These supports include integral flashing to prevent water entry. Proper weathersealing of all roof penetrations is essential.

5.0.0 SAFETY CONSIDERATIONS IN PV SYSTEMS

PV system installation often involves working in extreme heat on slanted, slippery roof surfaces ten or more feet above the ground (*Figure 23*). Always wear sunscreen, sunglasses, and non-slip footwear. Use fall protection in the form of guardrails. If guardrails cannot be installed, use a personal fall arrest system (PFAS). Ensure that straight ladders are positioned at an angle of one-fourth of the working height and extend beyond the roof surface by 36 inches. Maintain control over all

tools, equipment, and system components so they do not fall or slide off the roof, possibly injuring workers below. Wear a hardhat at all times. Avoid working in wet conditions. Never work where you can see lightning or hear thunder. Be careful when handling panels and other flat materials in windy conditions.

Never look directly into the sun. Avoid excessive sun exposure and stay hydrated. Drink eight ounces of water for every fifteen minutes spent working in the heat, as failure to stay hydrated can result in heat exhaustion or heat stroke. Do not use caffeinated drinks or alcohol before or during work.

In addition to working at elevation, you will be handling large, sharp, unwieldy PV panels that have the capacity to become energized the moment light strikes them (*Figure 24*). Plug the leads together to avoid the possibility of shock. Keep panels covered per the manufacturer's instructions so they remain de-energized. Wear protective gloves and seek assistance when lifting to avoid back strain.

PV panels may be hoisted onto a scaffold or moved using a material handling system. Scaffolds may only be erected and inspected by qualified individuals. All scaffold workers must be trained in the safe use of this equipment. Follow all safety procedures when using mechanical moving equipment. Never operate any equipment unless you are fully qualified.

Working with PV systems also involves working in hot attic areas. Provide ventilation where possible and avoid contact with insulation. Identify

Figure 23   PV array installed on a terra cotta roof.

Figure 24   Installing PV panels.

NCCER — *Contren® Learning Series*   57101-11

exit routes before working in tight areas. Exercise caution when working in attic spaces to avoid falling through the ceiling below. Install wood across ceiling joists to provide secure footing.

5.1.0 Fall Protection

The most common safety hazard when installing PV systems on an elevated structure is falls. This is compounded by the fact that there is usually no overhead attachment point, which is the approved method for protection. OSHA standards require PFAS to control falls for any working surface six feet or more above the ground. If you are forced to attach below the D-ring on your harness, you will fall that distance plus the six-foot length of a standard lanyard. This could result in a fall of up to 12 feet. A PFAS is not designed to meet those forces and will transfer much higher impacts when it arrests the fall. To combat this, the American National Standards Institute (ANSI) has changed the equipment manufacturing standards for fall protection. The new standard is commonly referred to as the 12-foot lanyard. The lanyard is still only six feet long, but the allowable deployment (extendable length) of the shock absorber has been increased. This lessens the force transferred to the body through the harness, reducing the likelihood of injury.

Before using a PFAS, you must examine the space below any potential fall point. Make sure it is clear of any hazards or obstructions. When assessing the required free length of a PFAS, you must account for the deployment of the shock absorber. In addition, PFAS manufacturers require up to a three-foot safety factor. Taking this safety factor into account, *Table 1* indicates the minimum distance required between any hazard/obstruction and the worker or attachment point.

As you can see, a typical single-story building may not provide enough height to use standard lanyards. Retractable lanyards that engage and stop falls in less than two feet are usually the best

choice for fall protection when used below 20 feet or on any type of lift equipment.

ANSI has also addressed the locking mechanism (gate) on snap hooks. OSHA requires the gate to be rated for 350 pounds when the rest of the system is rated at 5,000 pounds. This creates a weak link in the PFAS. ANSI issued a new standard increasing the strength of these gates to 3,600 pounds.

> **NOTE**
> Ladders, scaffolds, stairs, and similar types of equipment have other fall protection standards. Make sure you know and follow your company safety policies when using this equipment.

5.2.0 Battery Hazards

Deep cycle batteries release hydrogen when they are overcharged. This gas can build up to unsafe levels if batteries are poorly vented. The pressure rises within the cells until a spark or short circuit ignites the gas and it explodes. Battery explosions can cause severe injury or death. Replace any

> **On Site**
>
> ## Battery Temperature
>
> Batteries will begin to show increased temperatures prior to failure. A battery temperature sensor can be used with certain charge controllers to monitor the batteries and shut down the system at excessive temperatures.
>
>
>
> 101SA03.EPS

Table 1 Required Free Lengths for Various Lanyards

Lanyard Type	Lanyard Length	Shock Absorber Length	Average Worker Height	Safety Factor	Required Free Length
Standard 6' lanyard with 3.5' shock absorber	6'	3.5'	6'	3'	18.5'
New ANSI 6' lanyard with 4' shock absorber	6'	4'	6'	3'	19'
New ANSI 12' lanyard with 5' shock absorber	6' (12' freefall)	5'	N/A (accounted for by lanyard)	3'	20'

battery that feels unusually warm, appears swollen, or leaks. When handling batteries, wear approved, chemical-rated safety equipment including eye and face protection, gloves, and aprons.

Placing a wire across the terminals of a battery is called shorting the battery. Some batteries will explode when shorted. Never short a battery.

5.3.0 Electrical Hazards

Before working on any electrical equipment, the equipment and all inputs/outputs must be placed into an electrically safe work condition. This means that all equipment must be de-energized, any stored electrical energy must be discharged, and the equipment must be locked out and tagged before any work can be performed. It must also be tested to ensure that it is in an electrically safe condition using a known working tester.

Remember that PV systems have multiple power sources. These include utility power, PV system power, and batteries (if used). All possible power sources must be locked out and tagged before working on the system.

PV panels are rated up to 600V. In actual use, a PV panel may experience conditions that result in more current and/or voltage than listed on the panel nameplate.

A PV panel generates electricity when exposed to sunlight. Keep the panel covered until ready for installation and use insulated tools. When the panel is removed from the shipping container, plug the leads together to avoid the possibility of shock. Never touch the terminals when handling a panel.

Some of the most common electrical hazards in PV systems include cracked or broken panels, faulty connections, broken ground wires, and undersized wiring. Be aware of all potential hazards and exercise caution during installation and troubleshooting.

5.4.0 Meter Safety

Before using any meter, ensure that it is rated for the given application. PV installations require the use of Category III/IV meters. These meters are rated for use at the higher voltages expected in these systems. Inspect the meter and leads for damage before use. Never use a damaged meter. Wear safety glasses when using meters. Zero the meter before use, then set the meter to the correct range and select AC or DC.

> **NOTE**
>
> Detailed safety procedures were covered in the *Electrical Safety* module earlier in your training. Review this material.

On Site

Backfeed

Solar PV systems may present a backfeed hazard when the grid is de-energized. Do not rely on the inverter to disconnect power when the grid is de-energized. Provide warning labels on all service equipment and always perform lockout/tagout procedures.

101SA04.EPS

6.0.0 SITE ASSESSMENT

A careful site assessment is critical to a successful PV system. It provides an opportunity to interview the customer regarding their budget and expectations. It gives you a chance to review the customer's current electrical loads and the potential energy savings provided by PV power. You will also assess the site for optimal array orientation and equipment location, as well as any potential problems, including roof condition and shading.

> **NOTE**
>
> Local building codes and/or homeowner associations may place limits on array installation. For example, some local codes require that PV systems be installed on light-colored roofs to maximize light reflection, while some homeowner associations prohibit array installation on the street-facing roof face. Always consult local codes and homeowner association agreements prior to completing a site assessment.

Before leaving to perform a site assessment, ensure that you have the essential tools of the trade. This includes all required personal protective equipment, including sunscreen, protective eyewear, fall protection, non-slip shoes, and a hardhat. You will also need a site survey checklist, camera, calculator, angle finder, compass, ladder, line, and Solar Pathfinder™ (or equivalent).

> **WARNING!**
> When performing a site assessment, always wear sun protection, protective sunglasses, and shoes with non-slip soles. Use the appropriate fall protection and ensure that straight ladders extend beyond the roof surface by 36". To avoid permanent eye damage, never look directly into the sun.

6.1.0 Customer Interview

The customer interview lays the foundation for the system design. To the customer, the primary consideration in system selection is most likely the cost. For example, an entry-level system may supply 640W, which is enough power to run energy-efficient lights, a small refrigerator, and a few small appliances when supplied with batteries to overcome the motor starting currents. In comparison, a 3,600W system may supply enough power to operate a small home, including some larger motor loads. However, it is likely to cost three times as much as an entry-level system. When discussing system expectations with a customer, it is important to determine both their power expectations and budget. Find out whether they are seeking 100 percent replacement for grid power or if they will be satisfied with a partial replacement instead. Determine if the customer requires battery backup for the system.

6.2.0 Power Consumption

The customer interview must also include a load estimate. Many utilities provide free energy audits that can help in determining usage patterns. At a minimum, have the customer provide recent utility bills. In addition to precise kWh use for the current period, many utility bills show energy use over the last twelve months in the form of a simple bar chart on the bill. Customers may also be able to access this data online at their utility provider's website. Whenever possible, have the customer collect this data prior to the site visit. Find the month showing the highest power use, and then divide this value by the number of days in that month to find the average daily load. Multiply by 1,000 to convert kWh to Wh. Adjust this amount to reflect the customer's desired PV replacement percentage for the grid-connected load.

If current usage information is unavailable, it must be calculated using equipment wattages and expected hours of use. List all the loads to be supplied by the PV system and note the daily hours of use for each item during the critical design month (the coldest month). Multiply this by the nameplate wattage of the equipment to determine the daily Wh requirements.

> **NOTE**
> Appliance wattages can usually be determined from the owner's manual or equipment nameplate. If an appliance is rated in amps, multiply the amperage by the operating voltage to determine the value in watts.

6.3.0 Roof Evaluation

Document the type, size, age, and condition of the roof, including the strength and accessibility of supporting trusses. Consider the potential for wind loads and severe weather. If a roof is likely to require replacement in the next ten years, it should be replaced before the installation of a PV system.

> **WARNING!**
> If the roof structure is not sound, inform the building owner. Either a different location must be considered or the installation must be delayed until the roof is repaired. Never attempt to install a PV array on an unsound structure.

Consider the potential load in reference to the available space for the array. If the roof space is inadequate, a ground-mounted system must be considered. For example, suppose you have a south-facing, obstruction-free roof area of 200 square feet and will be installing polycrystalline PV panels. These panels have an average output of 10 watts per square foot. Multiply the available roof area by the wattage per square foot to find the largest system that can be installed:

200 sq ft × 10W/sq ft = 2,000W or 2.0kWh

This roof is large enough for a 2.0kWh system using typical polycrystalline panels. Note that actual panel outputs can vary widely and this is a general sizing strategy that can be applied to determine whether the roof has enough space for the installation of an array.

Sketch the south-facing roof area, including gables, vents, and other protrusions. Any shading caused by these protrusions will severely reduce the output of the PV array. If protrusions are noted, document an alternative location for the array. Next, determine the roof slope using an angle finder. Photograph the roof and the existing service.

> **NOTE**
>
> If the roof has terra cotta tile, the array installation must be performed by a specialized roofing contractor.

6.4.0 Array Orientation

Solar arrays work best when facing true solar south. True solar south is slightly different than a magnetic reference or compass south. A quick way to determine true solar south is to measure the length of time between sunrise and sunset, and then divide by two. The position of the sun at the resulting time represents true solar south.

> **NOTE**
>
> Solar south can also be determined using the National Oceanic and Atmospheric Administration Sunrise/Sunset Calculator at www.srrb.noaa.gov.

PV systems are designed to maximize output by optimal placement in relation to the motion of the sun. The most important values include *azimuth*, *altitude*, and *declination*.

- *Azimuth* – For a fixed PV array, the azimuth angle is the angle clockwise from true north that the PV array faces. The default azimuth angle is 180° (south-facing) for locations in the northern hemisphere and 0° (north-facing) for locations in the southern hemisphere. This value maximizes energy production. In the northern hemisphere, increasing the azimuth favors afternoon energy production, while decreasing it favors morning energy production. The opposite is true for the southern hemisphere.
- *Altitude* – The altitude is the angle at which the sun is hitting the array. In many areas, panels are simply set to an angle that matches the local latitude. This provides a yearly average maximum output power. However, seasonal adjustments can increase the output in locations with high irradiance. In small systems, panels can be manually adjusted to optimize the tilt angle. Large systems are more likely to use automatic

tilt adjustment. General recommendations for tilt angle include the following:
- *Maximum output power in winter* – In the northern hemisphere during the winter

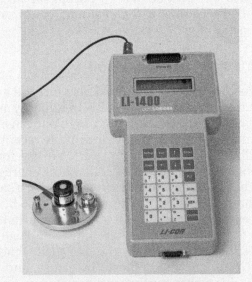

months, the tilt angle should equal local latitude +15°.

- *Maximum output power in summer* – In the northern hemisphere during the summer months, the tilt angle should equal local latitude –15°.
- *Seasonal adjustment* – In March and September, the tilt angle equals the local latitude. In June, the tilt angle equals latitude –15°, and in December, the tilt angle equals latitude +15°. These adjustments ensure maximum efficiency at all times.

• *Declination* – The declination of the sun is the angle between the equator and the rays of the sun. It ranges between +23.45° on the summer solstice (on or about June 21) to –23.45° on the winter solstice (on or about December 21). During the spring equinox (on or about March 21) and the fall equinox (on or about September 21), the angle is zero. Because declination changes throughout the year, the optimal panel tilt angle also changes. This seasonal tilt of the Earth causes longer shadows in winter because the sun is lower in the sky.

The Solar Pathfinder™ is a manual tool used to provide a full year of solar data for a specific location (*Figure 25*). It includes sun path diagrams for various latitude bands and an angle estimator for determining the sun's altitude and azimuth at various times of year. To use this instrument, go to the National Geophysical Data Center website at www.ngdc.noaa.gov and follow the links to find the latitude and declination angle for the desired location. Use the latitude to select the correct band diagram and a compass to align the unit to the desired angle of declination. Next, document shading by tracing the trees in the reflection using a grease pencil. You can also take a picture of the reflection rather than tracing it. The PV system must be installed outside of the shaded area or trees must be removed to eliminate shading. Companion software is available that generates monthly sun paths for each specific site latitude instead of using the latitude band diagrams.

The Solmetric SunEye™ is an electronic device that allows users to instantly assess the total potential solar energy of a site (*Figure 26*). This device shows site shading in a digital display rather than through tracings. It is more expensive than the Solar Pathfinder™ but has the advantage in ease of use.

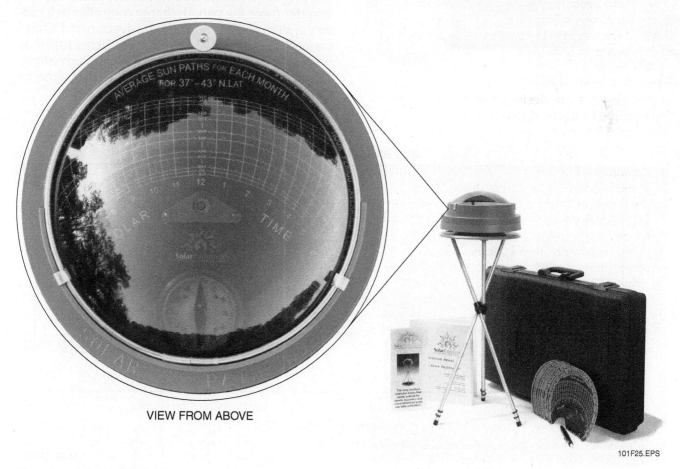

VIEW FROM ABOVE

101F25.EPS

Figure 25   Solar Pathfinder™.

 Introduction to Solar Photovoltaics

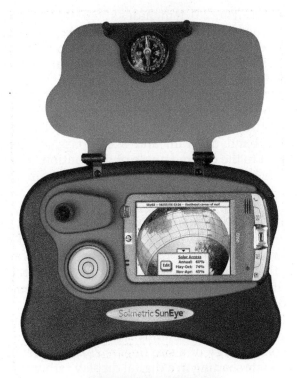

In most locations in the United States, winter produces the least amount of sunlight due to shorter days, increased cloud cover, and the sun's lower position in the sky. Insolation is usually highest in June and July and lowest in December and January. When selecting a site, choose an area that is free of shade between 9 AM and 3 PM on December 21st. *Figure 27* shows the path of the sun during the year for a specific location.

> **NOTE**
> You may need to adjust for Daylight Savings Time. See the instructions provided for the instrument in use.

For example, to find the approximate altitude and azimuth of the sun in April, follow along on the curve that reads "Apr. & Aug. 21" until you find 2:30 PM. The altitude is 45° while the azimuth is 60°. Note that values outside of the lines must be estimated.

6.5.0 Equipment Location

Sketch the location of the service in relation to the proposed array. Determine a location for the inverter (near the panel), charge controller (if a battery system will be installed), AC disconnect, DC disconnect, and ground. Group these items together whenever possible. Document a suggested battery location. The batteries should be located close to

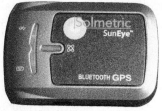

101F26.EPS

Figure 26   Solmetric SunEye™.

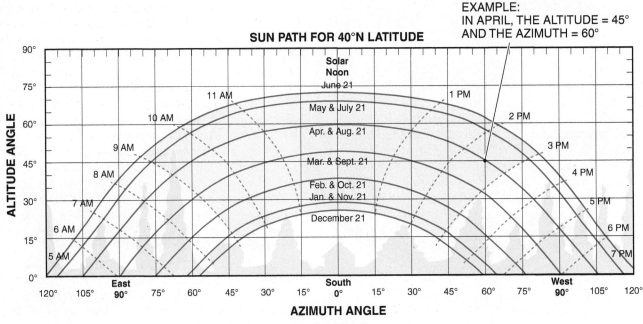

Figure 27   Sun path diagram.

NCCER — *Contren® Learning Series*   57101-11

the array to minimize voltage losses, but must be protected against temperature extremes. FLA batteries must be located in a vented compartment.

A typical site survey checklist is included in the *Appendix*.

7.0.0 SYSTEM DESIGN

PV system design begins with the load data collected during the site assessment. Next, various system choices must be made, such as sizing the PV panels, inverter, batteries, charge controller, and BOS components. All components must be properly matched to ensure efficient output.

7.1.0 Panel Nameplate Data

The panel nameplate contains a wealth of information that can be used in system sizing and design. It lists the model number, power output (P_{MAX}), maximum voltage and current at P_{MAX}, short-circuit current (I_{SC}), and open-circuit voltage (V_{OC}). The I_{SC} and V_{OC} are critical values when performing system calculations. The nameplate also lists the normal operating cell temperature (NOCT) at STC conditions, panel weight, maximum system voltage, fuse rating, and recommended conductor. See *Figure 28*. The minimum wire size listed on this panel nameplate is 12 AWG rated for 90°C. Type USE-2 cable is commonly used.

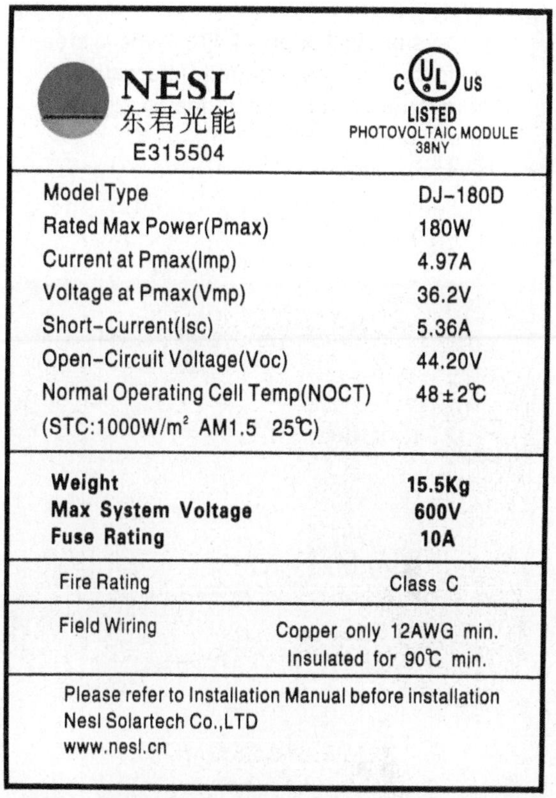

> **NOTE**
> The panel shown in *Figure 28* has a Class C Fire Rating and must be installed over a roof of equivalent fire resistance.

7.2.0 Solar Array Sizing

PV system sizing starts with the load and works backwards. To find the correct array size, divide the daily load in watt-hours determined in the site assessment by the number of sun hours per day, adjusted for the desired replacement percentage of grid power. For example, suppose a customer lives in an area with four peak sun hours per day, uses an average of 30kWh per day (30,000Wh/day) during the critical design month, and desires 50 percent replacement of grid power. The size of this PV array is calculated as follows:

30,000Wh/day ÷ 4 peak sun hours/day = 7,500Wh/day × 0.5.0 = 3,750Wh/day

The panel's I_{SC} rating must be multiplied by a factor of 1.25 when determining component ratings. *NEC Section 690.8* requires an additional multiplying factor of 1.25 for conductor and fuse

NESL 东君光能 E315504	c(UL)us LISTED PHOTOVOLTAIC MODULE 38NY
Model Type	DJ–180D
Rated Max Power(Pmax)	180W
Current at Pmax(Imp)	4.97A
Voltage at Pmax(Vmp)	36.2V
Short–Current(Isc)	5.36A
Open–Circuit Voltage(Voc)	44.20V
Normal Operating Cell Temp(NOCT)	48±2℃
(STC:1000W/m² AM1.5 25℃)	
Weight	**15.5Kg**
Max System Voltage	**600V**
Fuse Rating	**10A**
Fire Rating	Class C
Field Wiring	Copper only 12AWG min. Insulated for 90℃ min.
Please refer to Installation Manual before installation Nesl Solartech Co.,LTD www.nesl.cn	

101F28.EPS

Figure 28 PV nameplate data.

sizing. Refer to *NEC Table 690.7* for voltage correction factors based on ambient air temperatures.

7.3.0 Inverter Selection

Inverters are rated for specific battery voltages as well as for continuous wattage and surge wattage. The continuous wattage represents the total power output the inverter can support over time. The surge wattage accounts for momentary inrush currents. To select the appropriate inverter size, add up the wattages of all loads that are likely to operate at the same time. This determines the minimum continuous wattage. Next, consider the potential surge of each load to determine the minimum surge wattage. This is typically 150 to 200 percent of the continuous rating. For example, a 1,500W inverter might have a 2,500W surge rating.

Inverters are also selected based on the application. Off-grid systems use inverters with integral battery backup, while grid-tied systems use standard inverters (*Figure 29*).

All inverters list their efficiency ratings in the product data. For example, a typical grid-tied 2,000W inverter might have an efficiency of 95 percent. Values near or slightly above 95 percent are common. The inverter efficiency must be taken into consideration when sizing the inverter for the application.

STANDARD THREE-PHASE INVERTER

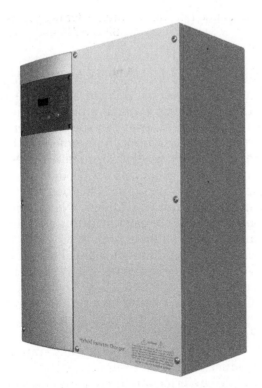

INVERTER WITH BATTERY SYSTEM

101F29.EPS

Figure 29 Inverters are selected based on the desired size and application.

7.4.0 Battery Bank Sizing

The size of the battery bank depends on the storage capacity required, the maximum charge and discharge rates, and the minimum temperature at which the batteries will be used. Wiring batteries in series increases the voltage. Wiring batteries in parallel increases the current (amp-hour) capacity, but does not affect the voltage. *Figure 30* shows a typical battery installation.

The amp-hour requirements of DC batteries must account for the voltage difference between the system and the batteries. For example, suppose you have a 24VDC nominal battery that is supplying a 4A, 120VAC load with a duty cycle of 4 hours per day. The load is 4A × 4 hours = 16Ah. In order to find the true battery load, divide the load voltage by the nominal battery voltage, and then multiply this times the amp-hours. In this case:

Battery capacity = (load voltage/nominal battery voltage) × Ah

Battery capacity = 120V/24V = 5 × 16Ah = 80Ah

Alternatively, you can determine the total required watt-hours, and then divide by the nominal DC voltage:

Battery capacity = 4A × 120VAC × 4 hours = 1,920 watt-hours/24VDC = 80Ah

Additional capacity can be added to supply the desired number of days of autonomy or battery backup. A typical system is sized for three days of autonomy. To calculate the battery bank size for

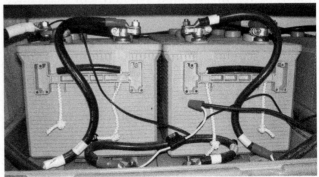

101F30.EPS

Figure 30 Installed batteries.

the desired days of autonomy, multiply the battery capacity by the number of days of required power times the desired DOD. For example, suppose you have a standalone system with a daily load of 80Ah at 24V, three days of autonomy, and a DOD of 80 percent. The size of the battery bank can be determined as follows:

$$\text{Battery bank} = \text{capacity} \times \text{days of autonomy} \times \%\text{DOD}$$

$$\text{Battery bank} = 80\text{Ah} \times 3 \text{ days} \times 80\% = 192\text{Ah}$$

This system could be served by eight batteries rated at 24V each (192Ah/24V = 8).

7.5.0 Selecting a Charge Controller

Properly sized charge controllers help to ensure smooth power and long battery life. The selection of a charge controller depends on the application. PV systems with outputs over 200W often use MPPT charge controllers. They are more expensive but worth the investment in greater efficiency when installed to support larger systems. Smaller systems typically use PWM charge controllers.

Charge controllers must be carefully matched to the size of the battery bank (normally 12V, 24V, or 48V), the AC short-circuit current (I_{SC}) of the solar array, and the maximum system voltage. To size a charge controller, multiply the I_{SC} of the entire array by a factor of 1.25. (Note that this is a minimum value and may vary depending on manufacturer.) The charge controller maximum input voltage must be higher than the maximum system voltage in PWM charge controllers. The design of MPPT charge controllers allows a greater nominal array voltage than the battery voltage. This permits the use of smaller wire between the battery and array.

7.6.0 Adjusting PV Conductors

PV system conductors must be adjusted for many factors. These include the following:

- Temperature
- Continuous duty
- Conduit fill (more than three conductors in a raceway)
- Voltage drop

7.6.1 Temperature Adjustment

Per *NEC Table 310.15(B)(3)(c)*, conduit exposed to sunlight on a rooftop requires an adjustment for the ambient temperature. This adjustment depends on the amount of spacing between the conduit and the roof surface.

Per *NEC Section 690.7(A)*, the maximum PV system voltage is calculated as the sum of the rated open-circuit voltages for all series-connected panels corrected for the lowest expected ambient temperature. The rated open-circuit voltage (V_{OC}) of a PV panel is measured at 25°C (77°F). Temperature correction is required to ensure that the output voltage will not exceed the maximum input voltage of the inverter or the rating of conductors, disconnects, and overcurrent devices. The voltage correction factors are listed in *NEC Table 690.7* and shown in *Table 2*.

The lowest available temperature for a specific area is sometimes known as the winter design temperature, and the month in which it occurs is known as the critical design month. These values

Table 2 Voltage Correction Factors for Crystalline and Multicrystalline Silicon Modules (Data from *NEC Table 690.7*)

Correction Factors for Ambient Temperatures Below 25°C (77°F). (Multiply the rated open circuit voltage by the appropriate correction factor shown below.)		
Ambient Temperature (°C)	**Factor**	**Ambient Temperature (°F)**
24 to 20	1.02	76 to 68
19 to 15	1.04	67 to 59
14 to 10	1.06	58 to 50
9 to 5	1.08	49 to 41
4 to 0	1.10	40 to 32
−1 to −5	1.12	31 to 23
−6 to −10	1.14	22 to 14
−11 to −15	1.16	13 to 5
−16 to −20	1.18	4 to −4
−21 to −25	1.20	−5 to −13
−26 to −30	1.21	−14 to −22
−31 to −35	1.23	−23 to −31
−36 to −40	1.25	−32 to −40

can be found using local weather information or data published by the American Society of Heating, Refrigerating, and Air-Conditioning Engineers (ASHRAE).

To use *Table 2*, multiply the factor for the lowest expected temperature by the rated V_{OC} found on the panel nameplate. For example, suppose a PV panel has a V_{OC} of 20V and four of these panels will be connected in series. This system will be installed in an area where the lowest expected temperature is 0°C (32°F). Per *Table 2*, the correction factor for this temperature is 1.1.0. The system output voltage can be calculated as follows:

$$\text{System } V_{OC} = \text{panel } V_{OC} \times \text{no. of panels} \times$$
$$\text{correction factor}$$

$$\text{System } V_{OC} = 20V \times 4 \times 1.10 = 88V$$

> **NOTE**
> *NEC Table 690.7* is based on crystalline PV panels that have been in use for many years. Newer panels may incorporate alternative semiconductor materials with varying temperature coefficients. Due to these variations, the manufacturer's requirements for temperature adjustment must be used where available per *NEC Section 690.7(A)*.

7.6.2 Continuous Duty

PV loads are considered continuous loads. Per *NEC Section 690.8(B)*, conductors used for continuous duty (operated more than three hours at a time) must be sized at 125 percent of the continuous load. For example, if the load is 20A, the conductor must be rated as follows:

$$\text{Continuous duty load} = \text{actual load} \times 125\%$$
$$\text{Continuous duty load} = 20A \times 125\% = 25A$$

7.6.3 Conduit Fill

Per *NEC Section 310.15(B)(3)(a)*, more than three conductors in a raceway or cable must be adjusted using *NEC Table 310.15(B)(3)(a)*. For example, suppose a 6 AWG USE-2 conductor is to be used with a total of four conductors in a raceway. Referring to *NEC Table 310.15(B)(3)(a)*, the deduction for 4 through 6 cables in a raceway is 80 percent of the values found in *NEC Table 310.15(B)(16)*. A 6 AWG USE-2 conductor is rated at 75A per *NEC Table 310.15(B)(16)*. The adjusted value can be calculated as follows:

$$\text{Adjusted amperage} = \textit{NEC Table 310.15(B)(16)}$$
$$\text{amperage} \times 80\%$$
$$\text{Adjusted amperage} = 75A \times 0.80 = 60A$$

7.6.4 Voltage Drop

Excessive circuit lengths can cause a significant decrease in system output voltages. This is particularly noticeable in low-voltage PV systems. In PV systems, voltage drop should be limited to 3 percent or less.

8.0.0 INSTALLATION

The two most common system installations are roof-mounted and ground-mounted arrays. Roof-mounted installations are more common in residential work where the array space on the ground may be limited due to lot size. In addition, roof-mounted arrays are more aesthetically pleasing and this location helps to limit accidental contact with energized components. Ground-mounted installations are often used for commercial systems where access to unauthorized personnel can be limited by restricted areas or fencing. A ground-mounted installation is more time-intensive than a roof installation due to the larger support system. The safe installation of either type of system requires an understanding of the forces exerted on panels and support structures.

8.1.0 Forces Exerted on the Panels/ Support System

Solar arrays are subjected to a variety of forces that can stress the panels and structural components. These forces include expansion and contraction, drag, and wind.

8.1.1 Expansion and Contraction

As a solar panel heats and cools, the metal frame and array mounting structure expand and contract. This change in length is multiplied when panels are racked side by side in a large array. For this reason, any installation in excess of 50 feet in length should include an expansion joint for both the array and the mounting structure or as designed by the professional engineer (PE) of record. The solar panel expansion joint should not straddle the mounting panel expansion joint.

8.1.2 Drag

If an array is installed in an area with heavy snow/ice loading, the solar panels will be subjected to a drag force that pulls the panels at a downward angle. To resist the drag force, the solar panel must have support on the lower edge of the panel frame. Additional mounting clips may be required for systems with excessive drag.

8.1.3 Wind

Wind loads can stress or damage PV panels and supports. In windy conditions, a solar array can exert several thousand pounds of uplift force against the structural supports. Wind loads are particularly critical with angled roof supports as the panels can act as a sail, pulling the array away from the structure.

Careful evaluation of wind loads is essential to ensure a safe installation and is also required to secure certain types of rebates. Wind loads are labeled on support systems and their associated fasteners. Some pre-engineered systems may be supplied with a PE stamp to indicate that they have been evaluated for a specific wind load. All installed systems require PE certification. In addition, systems installed in earthquake-prone areas must be rated to resist seismic loads.

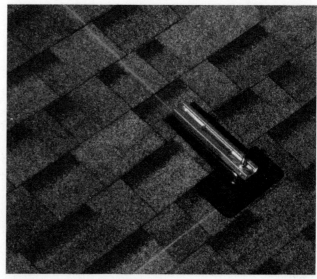

101F31.EPS

Figure 31   Roof layout.

> **NOTE**
>
> Contact the local utility to coordinate the installation of the special metering equipment required with grid-tied and grid-interactive PV systems.

> **WARNING!**
>
> Only qualified individuals may install PV system arrays. Wear all required personal protective equipment when installing PV systems. Apply sunscreen and stay hydrated when working in hot temperatures. Always use fall protection on work surfaces six feet or more above the ground. Do not step on or rest objects on solar panels. Be aware of all electrical hazards.

> **CAUTION**
>
> All roof penetrations, including fastener heads, must be weathersealed to avoid leaks.

8.2.0 Roof-Mounted Installations

An approved fire-rated roofing product must be installed on the roof deck before installation of a PV array. Selection of the roofing product depends on local code requirements, building design, and user preference. In many cases, the fire-rated product consists of asphalt shingles or concrete roof tiles. It must cover the entire roof surface, including areas where PV panels will be installed. In order to minimize shading losses at low sun angles, follow the manufacturer's requirements for panel spacing. For best appearance, center the array on the roof. The roof layout is determined using chalk lines, as shown in *Figure 31*. Roof panels can be laid out in either a portrait (vertical) or landscape (horizontal) orientation.

All mounting holes must be predrilled. To ensure system integrity, all mounts must be attached to structural members such as rafters and trusses. Follow the manufacturer's instructions for tightening fasteners to the proper torque. Secure connections are essential because the wind force and other loads are transferred to the roof structure via the mounting rails. Rails are normally fastened using stainless steel lag bolts since many roofs do not have attic spaces that would allow for access in order to use other types of fasteners. Wind loading can be minimized by mounting arrays no closer than three feet to any edge of the roof surface. Test each panel before installation.

Panels are secured to the roof deck in one of three ways: direct mount, rack mount, and stand-off mount:

- *Direct mount* – Direct mount systems mount directly to the roof. These systems provide an unobtrusive appearance and are subjected to the lowest wind loads. However, they prohibit air circulation beneath the panel. This increases the panel temperature, decreasing the output. *Figure 32* shows a direct-mount roof array laid out in a portrait orientation.
- *Rack mounts* – Rack mounts secure the panels using a triangular support. They may be supplied with or without an adjustable tilt angle.

101F32.EPS

Figure 32 Installed roof array.

101F33.EPS

Figure 33 Setting footers.

They are commonly used on flat roofs and designed to match the desired tilt angle. These mounts provide the greatest degree of air circulation and also provide easy access to electrical connections. However, they are also subject to the greatest degree of wind loads. When installing rack-mounted arrays, it is essential to follow the manufacturer's instructions regarding inter-row spacing. Improperly spaced arrays will create shadows, seriously reducing or even preventing system output voltages.

- *Standoff mounts*– Standoff mounts are the most common type of mounting system for sloped roofs. They provide three to five inches of space between the panel and the roof. This allows for a reasonable degree of air circulation while minimizing wind loads.

101F34.EPS

Figure 34 Installing the support system.

> **NOTE**
> Some commercial rooftop systems use a ballast (weighted) system rather than traditional roof mounts. Due to weight and wind loading, these systems must be engineered to match the specific structure and location.

8.3.0 Ground-Mounted Installation

When installing a ground-mounted array, the site must first be cleared of any vegetation and leveled. Next, concrete footers are installed to support the array (*Figure 33*). This anchor system requires engineering assistance to ensure sufficient protection against expected loads.

After the footers are installed, the support system is erected according to the manufacturer's instructions (*Figure 34*). *Figure 35* shows a completed support system.

101F35.EPS

Figure 35 Completed support system installation.

When the support system is complete, the panels are installed. Test each panel before installation. All fasteners must be tightened to the manufacturer's specifications (*Figure 36*). *Figure 37* shows a

Figure 36   Torquing fasteners.

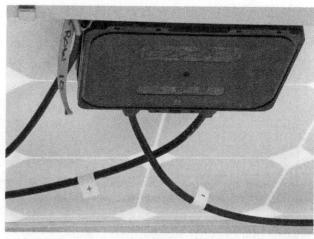

Figure 38   Making electrical connections.

Figure 37   Installed ground-mounted array.

completed ground-mounted array. White marble chips or other light-colored stone is used to reflect light and maximize panel output.

8.4.0 Electrical System Installation

After the array has been installed, the final electrical connections are made. The panels can be joined using traditional conductors and connectors, but are more often installed using quick-connect systems. The panels shown in *Figure 38* use a quick-connect system with sealed termination boxes. The factory wiring is identified for the positive (+) and negative (–) panel output polarities.

Size and install the equipment grounding conductor in accordance with local requirements and the *NEC®*. Attach the equipment grounding conductor to the panel frame using the hole and hardware provided.

WARNING!

Always observe proper precautions when connecting or disconnecting panels exposed to light since hazardous voltage may be present. Note that quick-connect termination boxes contain no user-serviceable components. Do not open the boxes or alter the wiring in this type of system.

Connections are also made to the inverter, batteries and charge controller (if used), panel, and disconnects. Use Type USE-2 copper wire (sunlight resistant and 90°C), for all wiring exposed to weather. The minimum wire size is listed on the panel nameplate. For example, a panel might list the minimum wire size as 12 AWG and the maximum wire size as 8 AWG. (Remember, wire sizes increase as the AWG size gets smaller, then they jump to 1/0, 2/0, and so on.)

Grid-tied systems typically feed into the last two spaces at the bottom in the main service panel. When adding PV power to a panel, do not exceed 120 percent of the bus rating. See *NEC Section 705.12(D)(2)*. Additional capacity may be added when necessary by reducing the size of main breaker or tapping ahead of the bus feed and adding a fused disconnect. Note that all components must be labeled per the *NEC®* and the proper working clearance must be provided for each piece of equipment. *Figure 39* shows the completed electrical installation.

After the system has been installed, it must be tested (commissioned) to ensure that it meets the expected outputs. Check for output voltage, current, and polarity. Program inverters and charge controllers. Review the operation of the system with the building owner and provide all operating manuals for the installed equipment.

 Introduction to Solar Photovoltaics

INVERTER

AC DISCONNECT

DC DISCONNECT

SURGE PROTECTOR

SUBPANEL

COMBINER BOX

101F39.EPS

Figure 39 Completed electrical installation.

8.5.0 Assessing System Output Power

Assessing the expected output of a grid-tied PV array requires the application of various panel correction factors. Typical correction factors include the following:

- Irradiance (varies between 1–10 percent)
- Array temperature (varies between 1–10 percent)
- Panel mismatch and production tolerances (4–5 percent)
- Dust (7 percent)
- Wiring losses (3 percent)
- Inverter conversion losses (4–5 percent)

Taking into account all of these system losses, the actual array output should be somewhere in the neighborhood of 70–80 percent of the STC value. This means that the measured output of a 1,000W PV system should be somewhere between 700W and 800W on a sunny day.

Assessing the expected output of a grid-interactive system is more complicated due to the addition of the batteries. Grid-interactive systems do not operate at maximum output power unless equipped with an MPPT charge controller. In addition, the batteries must be fully charged and the system must allow full input power to the inverter in order to assess system operation using this method. Many charge controllers and inverters provide system output and efficiency data on a digital display.

9.0.0 MAINTENANCE

While PV systems are generally low on maintenance requirements, there are certain checks and tests that must be performed to ensure proper system performance. The most important factor in the continuing operation of a well-designed system is periodic maintenance. Common quarterly maintenance checks include the following:

> **WARNING!**
>
> Only qualified individuals may perform PV system maintenance tasks. Wear all required personal protective equipment when installing PV systems. Apply sunscreen and stay hydrated when working in hot temperatures. Always use fall protection on work surfaces six feet or more above the ground. Do not step on or rest objects on solar panels. Be aware of all electrical hazards.

- Examine the panels for dirt, dust, sap, or bird droppings and clean using the manufacturer's recommended solution.
- Check panel glass for breakage or cracks.

- Check for loosened, corroded, or burnt electrical connections.
- Observe the condition of all wiring insulation.
- Make sure all grounding clips and wires are intact and secure. Inspect and test the grounding system.
- Check for loosened fasteners. Ensure that all fasteners are tightened to the manufacturer's torque requirements.
- Observe the growth of trees and shrubbery in the vicinity of the array. Have the building owner remove or trim any vegetation that is in danger of shading the array.
- Check the operation of inverters, batteries, and charge controllers. These components are normally replaced one or more times during the life of the system.

10.0.0 TROUBLESHOOTING

PV systems are subject to various environmental stresses, including temperature changes, wind, dirt, dust, lightning, and static electricity. In addition, manufacturing imperfections, improper installation, and equipment age can lead to component or system failures. A variety of tests can be performed to identify and isolate system malfunctions.

10.1.0 Loose or Corroded System Connections

Panel malfunctions can often be traced to a loose or corroded system connection. One way to locate a malfunctioning panel in an array is to shade a small portion of each panel in turn using a piece of cardboard or similar obstruction. This is known as a selective shading test. With the array connected and working, monitor the current. Now, shade a portion of one panel. You should see the current drop. If the current does not drop, then the panel that you are shading is out of the circuit. Look for a fault in the wiring of that panel, or of another panel that is wired in series with it.

10.2.0 Inverter Losses

Inverter losses are often caused by dirty panels. If you observe a drop in the inverter output and it is not due to shading, check the panels for dirt/dust

---

Think About It

Fastener Checks

Some installers use permanent marker to draw a line over the center of the fastener when it is initially installed and after it has been checked for proper torque. That way, they can tell at a glance when fasteners have become loose. Can you find the loosened fastener in this picture?

101SA06.EPS

 Introduction to Solar Photovoltaics

buildup and clean if necessary. Always use the manufacturer's recommended cleaning solution.

10.3.0 Heat Fade

Heat fade can be observed if the system operates inefficiently during periods of high heat. It is usually caused by poor connections or undersized wiring. Heat fade can be confirmed by throwing water on the panels to cool them down while monitoring current fluctuations. Heat fade is more common in grid-interactive and standalone systems due to the addition of the battery system.

10.4.0 Burnt Terminals

Over time, repeated temperature cycling, oxidation, and corrosion may eventually cause enough resistance to burn terminal connections. When repairing burnt terminals, replace all metal parts that have been severely oxidized.

Burnt terminals are usually the result of wiring too many panels in the same circuit. If this is the case, rewire some of the panels on a second circuit.

10.5.0 Bypass Diode Failure

Most PV panels have bypass diodes in the termination boxes to protect cells from overheating during sustained periods of partial shading. Bypass diodes are wired in parallel to prevent the reverse bias voltage caused by partial shading. Diodes may occasionally fail due to lightning or other voltage surges. If this occurs, the diode will normally short out and drop the panel voltage. Replace it per the manufacturer's instructions.

11.0.0 CODES AND STANDARDS

Industry codes and standards are designed to monitor, review, and enforce safety policies and procedures. The PV industry is covered under the following codes and standards:

- Local and national building codes
- International Association of Plumbing and Mechanical Officials (IAPMO): *Uniform Solar Energy Code*

On Site

Microinverters

Enphase microinverters attach directly to individual solar panels using low-voltage DC wiring. This increases system safety and allows for easy expansion.

- Institute of Electrical and Electronics Engineers (IEEE): *IEEE 1547, Standard for Interconnecting Distributed Resources with Electric Power Systems*
- Underwriters Laboratories (UL): *UL Standard 1703, UL Standard for Safety, Flat-Plate Photovoltaic Modules and Panels* and *UL Standard 1741, Standard for Inverters, Converters, Controllers and Interconnection System Equipment for Use with Distributed Energy Resources*
- Occupational Safety and Health Administration (OSHA): *OSHA Standard 1910.302, Electric Utilization Systems*
- National Fire Protection Association (NFPA): *National Electrical Code®* (*NFPA 70*) and *Standard for Electrical Safety in the Workplace®* (*NFPA 70E*)

12.0.0 EMERGING TECHNOLOGIES

There are tremendous research efforts in the field of PV systems. Some of the emerging areas of research include replacements for traditional semiconductors, while others involve methods of improving efficiency in solar collection and energy storage.

One new type of system replaces silicon with a light-sensitive dye that absorbs light and produces current. These are known as electrochemical solar cells. When this technology matures, electrochemical cells will be very popular as they are simple and less expensive than traditional solar cells.

Another developing technology uses solar concentrator systems. Solar concentrators use mirrors or lenses to focus light onto specially designed cells. Unlike other PV systems, concentrator systems will not operate under cloudy conditions. They generally follow the sun's path through the sky during the day using single-axis tracking. Single-axis tracking adjusts the vertical tilt of the panels. To adjust to the sun's varying height in the sky through the seasons, dual-axis tracking is sometimes used. Dual-axis tracking adjusts both the horizontal and vertical axes of the panels.

Research in solar energy storage includes the development of household fuel cells to replace traditional battery systems. These differ from traditional fuel cells, which use rare components and are very expensive to manufacture. This process uses sunlight and a special chemical component known as a catalyst to split water into hydrogen and oxygen, which are later recombined in the fuel cell to produce energy. It is clean, efficient, and inexpensive. According to researchers at the Massachusetts Institute of Technology (MIT), it is possible that household fuel cells could begin to replace wire-based energy delivery within the next decade.

SUMMARY

PV power provides a clean, renewable alternative to fossil fuels. It can be harnessed without disrupting the environment and produces no hazardous waste or emissions. It is also quiet, reliable, and requires little maintenance.

Solar cells produce electricity through the use of semiconductors. Many cells are contained in a solar panel, and these panels are electrically and mechanically connected in an array. The current produced by the array is DC current, which is converted to AC current by an inverter. It then travels to the main panelboard, where it is distributed to power lights and appliances in the structure.

Solar energy can be classified into four types of systems: standalone systems, grid-tied systems, grid-interactive systems, and utility-scale solar generating systems. Most residential and commercial customers use grid-tied and grid-interactive systems. These systems provide an advantage in that excess power can be sold back to the utility in the form of credits. In addition, utility and government rebates are available to offset the cost of installing these systems.

In standalone and grid-interactive systems, backup power is provided by a system of one or more batteries. These batteries are matched to the appropriate charge controller to provide grid-quality power.

As this technology continues to develop, it is likely that PV systems will become one of the primary sources of electrical power.

Review Questions

1. Which of the following is an active form of solar power?
   a. designing a building to maximize sunlight exposure in winter
   b. hanging clothes on a line
   c. heating water in an outdoor tank
   d. using a solar-powered calculator

2. When excess energy is sent back into the grid, it is known as _____.
   a. buyback metering
   b. grid metering
   c. net metering
   d. carryover metering

3. The use of solar power is increasing by _____.
   a. 5 percent per year
   b. 10 percent per year
   c. 15 percent per year
   d. 25 percent per year

4. Which of the following is an advantage of a grid-tied system?
   a. Excess energy can be sold back to the utility.
   b. It can provide PV power at night.
   c. It can operate independent of the utility grid.
   d. It provides power when there is a grid outage.

5. A solar boiler is a type of _____.
   a. grid-tied system
   b. grid-interactive system
   c. solar-generating system
   d. standalone system

6. When a motorized system is used to adjust an array to follow the motion of the sun, it is known as _____.
   a. trailing
   b. trolling
   c. tracing
   d. tracking

7. What is the current when the voltage is 120V and the resistance is 10 ohms?
   a. 10 amps
   b. 12 amps
   c. 120 amps
   d. 1,200 amps

8. Which of the following is true regarding series circuits?
   a. If the circuit is open at any point in a series circuit, current will still continue to flow.
   b. A series circuit provides multiple paths for current flow.
   c. A series circuit is a voltage multiplier.
   d. The resistance of the circuit is equal to the sum of the individual resistances.

9. One hp is equal to _____.
   a. 1 watt
   b. 76 watts
   c. 144 watts
   d. 746 watts

10. If an electric heater uses 1,200W for 10 hours, the power consumed is _____.
    a. 1.2kWh
    b. 12kWh
    c. 1,200kWh
    d. 12,000kWh

11. The prefix *kilo* means _____.
    a. 10
    b. 100
    c. 1,000
    d. 1,000,000

12. Wiring two 30V/5A panels in series produces _____.
    a. 30V/5A
    b. 30V/10A
    c. 60V/5A
    d. 60V/10A

13. Wiring solar panels in parallel _____.
    a. increases the voltage but does not affect the amperage
    b. decreases the voltage but does not affect the amperage
    c. increases the amperage but does not affect the voltage
    d. increases both the voltage and the amperage

14. Which of the following locations is likely to have the highest irradiance?
    a. A desert area with significant dust but few cloudy days
    b. A mountaintop with cool temperatures but no dust and few cloudy days
    c. A hot southern location at sea level
    d. A cool but cloudy northern location

15. The *NEC*® requirements for PV installations can be found in _____.
    a. *NEC Article 422*
    b. *NEC Article 450*
    c. *NEC Article 517*
    d. *NEC Article 690*

16. A type of PV panel having a characteristic flaked appearance is probably made up of _____.
    a. monocrystalline cells
    b. megacrystalline cells
    c. polycrystalline cells
    d. amorphous cells

17. The *NEC*® requirements for stationary battery installations can be found in _____.
    a. *NEC Article 480*
    b. *NEC Article 490*
    c. *NEC Article 551*
    d. *NEC Article 625*

18. An advantage in choosing FLA batteries over AGM batteries is that FLA batteries _____.
    a. are less expensive
    b. require no maintenance
    c. do not leak
    d. do not have special venting requirements

19. The best charge controller for use in a cold climate is a _____.
    a. MPPT charge controller
    b. PWM charge controller
    c. shunt charge controller
    d. relay transistor charge controller

20. Each piece of equipment in a PV system requires a disconnect per _____.
    a. *NEC Section 690.1*
    b. *NEC Section 690.7*
    c. *NEC Section 690.15*
    d. *NEC Section 690.41*

21. PV panels are rated up to _____.
    a. 120V
    b. 240V
    c. 480V
    d. 600V

22. PV systems require the use of _____.
    a. Cat I meters
    b. Cat II meters
    c. Cat II or III meters
    d. Cat III or IV meters

23. The total electrical load of a small hunting cabin is most likely to be served by a _____.
    a. 640W system
    b. 1,000W system
    c. 3,600W system
    d. 12,000W system

24. Suppose the sun rises at 6:00 AM and sets at 9:00 PM. What time should you note the position of the sun in order to determine true solar south?
    a. 6:00 AM
    b. 12:00 PM
    c. 1:30 PM
    d. 3:00 PM

25. The default azimuth angle for locations in the southern hemisphere is _____.
    a. 0 degrees
    b. 90 degrees
    c. 180 degrees
    d. 270 degrees

26. During the spring equinox, the angle of declination is _____.
    a. zero
    b. +23.45 degrees
    c. −23.45 degrees
    d. 90 degrees

27. Suppose a customer lives in an area with four peak sun hours per day, uses an average of 36kWh per day during the critical design month, and desires 50 percent replacement of grid power. What is the required system size?
    a. 1,500Wh/day
    b. 2,500Wh/day
    c. 3,500Wh/day
    d. 4,500Wh/day

28. Suppose you have a standalone system with a daily load of 100 Ah, two days of autonomy, and a DOD of 80 percent. What is the size of the battery bank?
    a. 100Ah
    b. 125Ah
    c. 160Ah
    d. 200Ah

29. An array's I_{SC} rating is 100A. The charge controller for this system must be rated at _____.
    a. 100A
    b. 125A
    c. 175A
    d. 200A

30. An expansion joint is required for any array longer than _____.
    a. 30 feet
    b. 40 feet
    c. 50 feet
    d. 60 feet

31. All installed PV systems require a PE stamp indicating the rated _____.
    a. expansion differential
    b. drag
    c. wind load
    d. snow load

32. A flat roof is most likely to use a _____.
    a. direct mount
    b. building-integrated system
    c. rack mount
    d. standoff mount

33. Which of the following is often the result of dirty panels?
    a. Heat fade
    b. Bypass diode failure
    c. Inverter losses
    d. Burnt terminals

34. Which of the following is often the result of wiring too many panels in the same circuit?
    a. Heat fade
    b. Bypass diode failure
    c. Inverter losses
    d. Burnt terminals

35. Which of the following is likely to use a dual-axis tracking system?
    a. a solar concentrator system
    b. a residential off-grid system
    c. a small commercial grid-tied system
    d. a residential grid-interactive system

Trade Terms Quiz

Fill in the blank with the correct trade term that you learned from your study of this module.

1. A unit of energy, usually of electrical energy, equal to the work performed by a single watt for one hour is called a(n) _____.

2. For a fixed PV array, the angle clockwise from true north that the PV array faces is its _____.

3. The measure of radiation density at a specific location is its _____.

4. A junction box used to connect strings of solar panels to create a larger array, and to provide a convenient array disconnect point is a(n) _____.

5. The material that exhibits the properties of both a conductor and an insulator is called a(n) _____.

6. A device that harnesses the energy produced by a chemical reaction between hydrogen and oxygen to produce direct current is a(n) _____.

7. A complete PV power-generating system including panels, inverter, batteries and charge controller (if used), support system, and wiring is called a(n) _____.

8. A rapid switching method used to simulate a waveform and provide smooth power control is a(n) _____.

9. A diode, often used to overcome partial shading, that directs current around a panel rather than through it is a(n) _____.

10. The measure of a location's relative height in reference to sea level is its _____.

11. The thickness of the atmosphere that solar radiation must pass through to reach the Earth is the _____.

12. A low-efficiency type of photovoltaic cell characterized by its ability to be used in flexible forms is considered to be _____.

13. Converting direct current to alternating current requires a(n) _____.

14. When current flows into the grid, it is called _____.

15. A PV cell or panel operating at a negative voltage, typically due to shading, is called a(n) _____.

16. A material to which specific impurities have been added to produce a positive or negative charge is said to be _____.

17. The battery charge controller that provides precise charge/discharge control over a wide range of temperatures is the _____.

18. The device used to regulate the charging and discharging of the battery system to prevent overcharge and excess discharge is the _____.

19. A type of PV cell formed by using thin slices of a single crystal and characterized by its high efficiency is said to be _____.

20. Standardized panel ratings based on a specific operating temperature, solar irradiance, and air mass are referred to as _____.

21. The sun's altitude and azimuth at various times of year for a specific location or latitude band is known as the _____.

22. To determine a location on the Earth in reference to the equator, find its _____.

23. The measure of the amount of charge removed from a battery system is its _____.

24. When a PV system operates inefficiently during periods of high heat, it is usually caused by poor connections or undersized wiring in a condition called _____.

25. A measure of the average height of the ocean's surface between low and high tide and used as a reference for all other elevations on Earth is _____.

26. During outages and after sundown, supplying a PV system with supplemental power that can function independently through a battery bank requires a _____.

27. Large solar farms designed to produce power in quantities large enough to operate a small city are called _____.

28. A grid-interactive system used with other energy sources, such as wind turbines or generators, is called a(n) _____.

29. A PV system built into the structure as a replacement for a building component such as roofing is called _____.

30. A device that maximizes the collection of solar energy by using mirrors or lenses to focus light onto specially designed cells is called a(n) _____.

31. By pouring liquid silicon into blocks and then slicing it into wafers to create nonuniform crystals with a flaked appearance, a type of PV cell called _____ is formed.

32. The equivalent number of hours per day when solar irradiance averages $1,000W/m^2$, also known as peak sun hours, is called _____.

33. The number of days a fully charged battery system can supply power to loads without recharging is its _____.

34. The angle at which the sun is hitting the array is called the _____.

35. A method of measuring power from the grid against PV power put into the grid is called _____.

36. An array-mounting system designed to adjust either the horizontal or the vertical axis of a panel to follow the movement of the sun is a(n) _____.

37. The panel support system, wiring, disconnects, and grounding that are installed to support a PV array is called the _____.

38. The air temperature of an environment is called the _____.

39. The PV system component consisting of numerous electrically and mechanically connected PV cells encased in a protective glass or laminate frame is called a _____.

40. A PV system that operates in parallel with the utility grid and provides supplemental power to the building or residence is called the _____.

41. A temporary decrease in grid output voltage typically caused by peak load demands is a(n) _____.

42. The position of a panel or array in reference to horizontal and often set to match local latitude or in higher-efficiency systems is called the _____.

43. The angle between the equator and the rays of the sun is called _____.

44. The distortion of light through Earth's atmosphere is called _____.

45. An array mounting system designed to adjust both the horizontal and vertical axes of a panel to precisely follow the movement of the sun is controlled by _____.

46. A semiconductor device that converts sunlight into direct current is called a(n) _____.

47. The type of PV cell that replaces silicon with a light-sensitive dye that absorbs light and produces current is a(n) _____.

48. A standalone PV system that typically provides power in remote areas and uses batteries for energy storage as well as battery-based inverter systems is a(n) _____.

49. The volunteer board of renewable energy system professionals that provides standardized testing and certification for PV system installers is the _____.

Trade Terms

Air mass
Altitude
Ambient temperature
Amorphous
Array
Autonomy
Azimuth
Backfeed
Balance of system (BOS)
Brownout
Building-integrated
  photovoltaics (BIPV)
Bypass diode
Charge controller
Combiner box
Concentrator

Declination
Depth of discharge (DOD)
Doped
Dual-axis tracking
Electrochemical
  solar cells
Elevation
Fuel cells
Grid-connected system
Grid-interactive system
Grid-tied system
Heat fade
Hybrid system
Insolation
Inverter
Irradiance

Latitude
Maximum power point
  tracking (MPPT)
Module
Monocrystalline
Net metering
North American Board
  of Certified Energy
  Practitioners (NABCEP)
Off-grid system
Peak sun hours
Photovoltaic (PV) cell
Polycrystalline
Pulse width-modulated
  (PWM)
Reverse bias

Sea level
Semiconductor
Single-axis tracking
Spectral distribution
Standalone system
Standard Test
  Conditions (STC)
Sun path
Thin film
Tilt angle
Utility-scale solar
  generating system
Watt-hours (Wh)

Nicolás Estévez

Chief Executive Officer
Neosolvis Engineering, C.S.P
Mayaguez, Puerto Rico

It's not such a big jump from Superman to solar energy, not to Nicolás Estévez. As a boy, he wanted to save the world wearing a cape, but today he wants to contribute by working on renewable energies, particularly solar. Nicolás believes it's the answer to many of societies' problems.

How did you get started in solar photovoltaics?
As a young person, I wanted to be Superman. Then, because I loved airplanes, I decided I wanted to be a pilot. (The best I did there is becoming a Frequent Flier.) But when I went to the university to study mechanical engineering, I became fascinated by thermodynamics and all things energy. My parents were very environmentally conscious, and though I was born in the US, I grew up in Venezuela and Puerto Rico. I became aware of the various problems society faces. So I did some soul-searching and found that I wanted to work on renewable energies, particularly solar thermal. It seems the best way for me to help society with the skills I've acquired.

Who inspired you to enter such a new field?
My parents. The general upbringing I got from them inspired me to seek out a career in something I feel passionate about and which contributes to the greater good. Solar engineering is what I realized this was. For several years, I was able to study and work in Europe, where renewable energy is a major priority—the number of solar panels on rooftops is amazing.

What do you enjoy most about your job?
Personal satisfaction that I'm doing good for society and the environment. And the variety of challenges, the technical nature, flexible hours, and travel.

Do you think training and education are important in construction?
Yes. It's important to be able to have a strong workforce that is up to date with current technologies and emerging markets.

Would you suggest construction as a career for others?
Yes, we need more people to build today's designs. The opportunities will be there for a very long time.

How do you define craftsmanship?
Just like I define character. Craftsmanship is doing your job to the best of your ability in a productive and safe manner, whether or not anyone will know.

Trade Terms Introduced in This Module

Air mass: The thickness of the atmosphere that solar radiation must pass through to reach the Earth.

Altitude: The angle at which the sun is hitting the array.

Ambient temperature: The air temperature of an environment.

Amorphous: A low-efficiency type of photovoltaic cell characterized by its ability to be used in flexible forms. Also known as thin film.

Array: A complete PV power-generating system including panels, inverter, batteries and charge controller (if used), support system, and wiring.

Autonomy: The number of days a fully charged battery system can supply power to loads without recharging.

Azimuth: For a fixed PV array, the azimuth angle is the angle clockwise from true north that the PV array faces.

Backfeed: When current flows into the grid.

Balance of system (BOS): The panel support system, wiring, disconnects, and grounding system that are installed to support a PV array.

Brownout: A temporary decrease in grid output voltage typically caused by peak load demands.

Building-integrated photovoltaics (BIPV): A PV system built into the structure as a replacement for a building component such as roofing.

Bypass diode: A diode used to direct current around a panel rather than through it. Bypass diodes are typically used to overcome partial shading.

Charge controller: A device used to regulate the charging and discharging of the battery system to prevent overcharge and excess discharge.

Combiner box: A junction box used to connect strings of solar panels to create a larger array, and to provide a convenient array disconnect point.

Concentrator: A device that maximizes the collection of solar energy by using mirrors or lenses to focus light onto specially designed cells.

Declination: The angle between the equator and the rays of the sun.

Depth of Discharge (DOD): A measure of the amount of charge removed from a battery system.

Doped: A material to which specific impurities have been added to produce a positive or negative charge.

Dual-axis tracking: An array mounting system designed to adjust both the horizontal and vertical axes of a panel to precisely follow the movement of the sun.

Electrochemical solar cells: A type of PV cell that replaces silicon with a light-sensitive dye that absorbs light and produces current.

Elevation: A measure of a location's relative height in reference to sea level.

Fuel cell: A device that harnesses the energy produced by a chemical reaction between hydrogen and oxygen to produce direct current.

Grid-connected system: A PV system that operates in parallel with the utility grid and provides supplemental power to the building or residence. Since they are tied to the utility, they only operate when grid power is available. Also known as a grid-tied system.

Grid-interactive system: A PV system that supplies supplemental power and can also function independently through the use of a battery bank that can supply power during outages and after sundown.

Grid-tied system: See *grid-connected system*.

Heat fade: A condition in which a PV system operates inefficiently during periods of high heat. Heat fade is usually caused by poor connections or undersized wiring.

Hybrid system: A grid-interactive system used with other energy sources, such as wind turbines or generators.

Insolation: The equivalent number of hours per day when solar irradiance averages $1,000W/m^2$. Also known as peak sun hours.

Inverter: A device used to convert direct current to alternating current.

Irradiance: A measure of radiation density at a specific location.

Latitude: A method of determining a location on the Earth in reference to the equator.

Maximum power point tracking (MPPT): A battery charge controller that provides precise charge/discharge control over a wide range of temperatures.

Module: A PV system component consisting of numerous electrically and mechanically connected PV cells encased in a protective glass or laminate frame. Also known as a PV panel.

Monocrystalline: A type of PV cell formed using thin slices of a single crystal and characterized by its high efficiency.

Net metering: A method of measuring power used from the grid against PV power put into the grid.

North American Board of Certified Energy Practitioners (NABCEP): A volunteer board of renewable energy system professionals that provides standardized testing and certification for PV system installers.

Off-grid system: A PV system typically used to provide power in remote areas. Off-grid systems use batteries for energy storage as well as battery-based inverter systems. Also known as a standalone system.

Peak sun hours: See insolation.

Photovoltaic (PV) cell: A semiconductor device that converts sunlight into direct current.

Polycrystalline: A type of PV cell formed by pouring liquid silicon into blocks and then slicing it into wafers. This creates non-uniform crystals with a flaked appearance that have a lower efficiency than monocrystalline cells.

Pulse width-modulated (PWM): A control that uses a rapid switching method to simulate a waveform and provide smooth power.

Reverse bias: A PV cell or panel operating at a negative voltage, typically due to shading.

Sea level: A measure of the average height of the ocean's surface between low and high tide. Sea level is used as a reference for all other elevations on Earth.

Semiconductor: A material that exhibits the properties of both a conductor and an insulator.

Single-axis tracking: An array mounting system designed to adjust either the horizontal or the vertical axis of a panel to follow the movement of the sun.

Spectral distribution: The distortion of light through Earth's atmosphere.

Standalone system: See *off-grid system*.

Standard Test Conditions (STC): Standardized panel ratings based on a specific operating temperature, solar irradiance, and air mass.

Sun path: The sun's altitude and azimuth at various times of year for a specific location or latitude band.

Thin film: See *amorphous*.

Tilt angle: The position of a panel or array in reference to horizontal. Often set to match local latitude or in higher-efficiency systems, the tilt angle may be adjusted by season or throughout the day.

Utility-scale solar generating system: Large solar farms designed to produce power in quantities large enough to operate a small city.

Watt-hours (Wh): A unit of energy typically used for metering.

Appendix

SITE SURVEY CHECKLIST

SITE SURVEY CHECKLIST		
☀ **GENERAL INFORMATION**		
Date of Survey:		
Site Name:		
Contact Name:		
Site Street Address:		
City: **State:** **Zip:** **Country:**		
Phone: ( ) **Fax:** ( )		
Email:		

1. ROOF OR OTHER ARRAY MOUNTING SURFACE

Check boxes or specify in the blank for items below.

1.01	**Type of Roof Material or Mounting Surface (Specify)**	
1.02	**Roof or Mounting Surface Condition**	
1.03	**Age**	
1.04	**Supporting Structure (e.g. roof trusses)**	
	☐ Accessible	
	☐ Adequate Strength	
1.05	**Roof or Mounting Surface Slope (e.g., 5/12, flat)**	
1.06	**Area (Sq. ft.)**	
	- Azimuth Direction (degrees E or W of true South)	
	- Eave Height (ft.)	
	- Ridge Height (ft.)	
1.07	**Accessibility to Proposed Array Location**	
	☐ Easy	
	☐ Moderate	
	☐ Unacceptable	
1.08	**Potential for Shading Proposed Array**	
	☐ None	
	☐ Slight	
	☐ Unacceptable	

2. INVERTER, UTILITY ACCESS, BATTERIES AND ENGINE-GENERATOR (AS APPLICABLE)

2.01	**Proposed Inverter Location (Specify)**	
2.02	**Accessibility to Proposed Inverter Location**	
	☐ Easy	
	☐ Moderate	
	☐ Unacceptable	
2.03	**Proposed Battery Location (Specify, if applicable)**	
2.04	**Accessibility to Proposed Battery Location**	
	☐ Adequate Ventilation	
	☐ Adequate Location	
	☐ Accessible	
2.05	**Proposed Engine-Generator Location (Specify, if applicable)**	
	☐ Adequate Ventilation	
	☐ Adequate Location	
	☐ Accessible	

Copyright © Florida Solar Energy Center

101A01A.EPS

☀ RECOMMENDATION

Check the appropriate box below.

☐ Approve site for system installation

☐ Do not approve site for system installation (If site not approved, specify reasons for rejection below:)

_____

_____

_____

_____

☀ SURVEY REVIEWER INFORMATION

Name:

Organization:

Signature: **Date:**

Please list other committee members reviewing this design:

Name	Organization

SKETCH ROOF AREA AND PROPOSED ARRAY LOCATION (OR ATTACH ON A SEPARATE PAGE)

Available Roof Area (sq. ft.)

101A01B.EPS

Additional Resources

This module presents thorough resources for task training. The following resource material is suggested for further study.

IEEE 1547, Standard for Interconnecting Distributed Resources with Electric Power Systems, Latest Edition. Los Alamitos, CA: Institute of Electrical and Electronics Engineers (IEEE).

National Electrical Code® (NFPA 70), Latest Edition. National Fire Protection Association (NFPA): Quincy, MA.

Occupational Safety and Health Standard 1910.302, Electric Utilization Systems, Latest Edition. Washington, DC: OSHA Department of Labor, U.S. Government Printing Office.

Photovoltaic Systems, Second Edition. James P. Dunlop. Orland Park, IL: American Technical Publishers.

Standard for Electrical Safety in the Workplace® (NFPA 70E), Latest Edition. National Fire Protection Association (NFPA): Quincy, MA.

UL Standard 1703, UL Standard for Safety, Flat-Plate Photovoltaic Modules and Panels, Latest Edition. Camas, WA: Underwriters Laboratories.

UL Standard 1741, Standard for Inverters, Converters, Controllers and Interconnection System Equipment for Use with Distributed Energy Resources, Latest Edition. Camas, WA: Underwriters Laboratories.

Uniform Solar Energy Code, Latest Edition. Ontario, CA: International Association of Plumbing and Mechanical Officials (IAPMO).

Figure Credits

© iStockphoto.com/Pgiam, Module opener

NASA, 101SA01

Mike Powers, 101F02, 101F16, 101F18–101F21, 101F24, 101SA04, 101F33–101F39, 101SA06

Courtesy of DOE/NREL, Credit – Pete Beverly, 101SA02

Antonio Vazquez, 101F03, 101F04, 101F22

Sharp USA, 101F05, 101F12, 101F13 (panel), 101F14 (thin-film solar panel), 101F23, 101F32

eSolar Inc., 101F06

Nellis Air Force Base, 101F07

Topaz Publications, Inc., 101F13 (inset), 101F28, 101F30, 101F31

Photo courtesy of Energy Conversion Devices, Inc. & United Solar Ovonic LLC, 101F14 (flexible thin-film)

Fronius International GmbH, 101F15

Outback Power Systems, 101F17

Schneider Electric, 101SA03, 101F29

The Eppley Laboratory, 101SA05 (mounted pyranometer)

Copyright LI-COR, Inc. and used by permission, 101SA05 (handheld pyranometer)

Solar Pathfinder, 101F25

With permission of Solmetric Corporation, 101F26

Table 2 reprinted with permission from *NFPA 70®, National Electrical Code®*, Copyright © 2010, National Fire Protection Association, Quincy, MA. This reprinted material is not the complete and final position of the NFPA on the referenced subject, which is represented only by the standard in its entirety.

Florida Solar Energy Center® (FSEC®), a research institute of the University of Central Florida, Appendix

Trainees with successful module completions may be eligible for credentialing through NCCER's National Registry. To learn more, go to **www.nccer.org** or contact us at **1.888.622.3720.** Our website has information on the latest product releases and training, as well as online versions of our Cornerstone magazine and Pearson's product catalog.

Your feedback is welcome. You may email your comments to **curriculum@nccer.org,** send general comments and inquiries to **info@nccer.org**, or use the User Update form at the back of this module.

 V.2 11/13

Site Assessment

57102-11

Trainees with successful module completions may be eligible for credentialing through NCCER's National Registry. To learn more, go to **www.nccer.org** or contact us at **1.888.622.3720**. Our website has information on the latest product releases and training, as well as online versions of our *Cornerstone* magazine and Pearson's product catalog.

Your feedback is welcome. You may email your comments to **curriculum@nccer.org**, send general comments and inquiries to **info@nccer.org**, or use the User Update form at the back of this module.

 V.2 11/13

Objectives

When you have completed this module, you will be able to do the following:

1. Determine customer needs:
   - Determine electrical load and energy use by review of utility bills, meter readings, measurements, and/or customer interviews.
   - Estimate and/or measure the peak load demand and average daily energy use for all connected loads.
2. Assess any site-specific safety hazards and/or installation considerations.
3. Identify and use the tools and equipment required for conducting site surveys for PV installations.
4. Identify, select, and sketch a suitable location for PV array installation, including proper orientation, sufficient area, adequate solar access, and structural integrity.
5. Select suitable locations for installing inverters, control(s), batteries, and other components.
   - Identify essential loads for battery systems.
   - Identify opportunities for the use of energy-efficient equipment/appliances, conservation, and energy management practices.
6. Acquire and interpret site solar radiation and temperature data to establish performance expectations and use in electrical system calculations.

Performance Tasks

Under the supervision of the instructor, you should be able to do the following:

1. Given the results of a customer interview and the sample house drawing provided, complete a site survey and checklist.

Trade Terms

Magnetic declination
Pitch

Prerequisites

Before you begin this module, it is recommended that you successfully complete *Core Curriculum* and *Solar Photovoltaic Systems Installer*, Module 57101-11. It is also suggested that you shall have successfully completed the following modules from the Electrical curriculum: *Electrical Level One*, Modules 26101 through 26111; *Electrical Level Two*, Modules 26201, 26205, 26206, and 26208 through 26211; *Electrical Level Three*, Modules 26301 and 26302; and *Electrical Level Four*, Modules 26403 and 26413.

Contents ───────────────────────────

Topics to be presented in this module include:

Figures and Tables

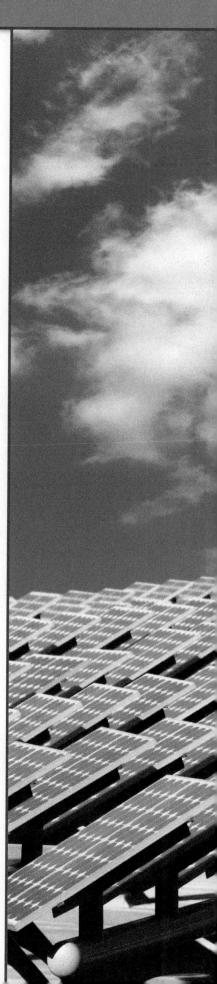

1.0.0 INTRODUCTION

A careful site assessment is essential to the safe and effective operation of a photovoltaic (PV) system. A site assessment is used to educate the customer about the potential benefits of a PV system and to gather a variety of information about the proposed site. This includes the following:

- Reviewing the customer's expectations and budget
- Analyzing the existing electrical loads, including making recommendations for additional energy-efficiency improvements
- Surveying potential locations for the PV array and balance-of-system (BOS) components
- Documenting site orientation and shading

After all of this information has been collected, it is presented to the system designer, who then selects the appropriate equipment and array configuration to meet both the customer expectations and the conditions of the site.

1.1.0 Assessment Tools and Equipment

Site assessment requires specific tools and equipment to access the installation location and document the site conditions. Ladders or aerial lifts are needed to access roofs. Shade analysis instruments are used to evaluate the best location for the panels. Generally speaking, the following tools are required at most locations:

- *Personal protection equipment* – This includes sunscreen, protective eyewear, fall protection, nonskid shoes, and a hardhat. Fall protection must be worn when working at heights of six feet or more. Remember that when assessing the conditions or taking measurements on an elevated surface such as a roof, tools must be controlled at all times to prevent them from falling and causing injury to anyone below.
- *Checklists* – When going out to assess a potential installation site, bring a checklist of tools and instruments needed for the job. Another important checklist is the site survey checklist. Checklists may vary by company, but typically include standard questions, such as owner contact information, type of system desired, and roof condition. The Florida Solar Energy Center offers a site survey checklist that can be accessed at **www.fsec.ucf.edu**.
- *Documentation tools* – These include a protractor, pencils, a ruler, graph paper, and a digital camera. Be sure to photograph and sketch all areas of the proposed installation site, including the existing electrical service and any obstructions and roof penetrations.

- *Distance measuring devices* – Reel-style tape measures may be needed when evaluating a large installation area. Another option is to use a measuring wheel (analog or digital). *Figure 1* shows a reel-style tape measure and a measuring wheel.

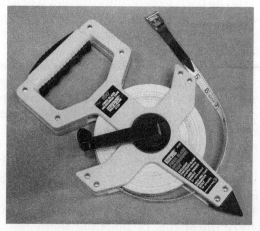

LONG REEL-STYLE MEASURE

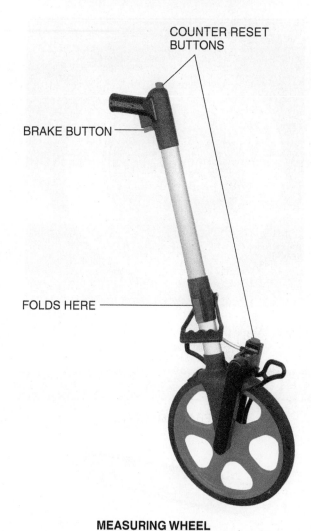

MEASURING WHEEL

102F01.EPS

Figure 1 Distance measuring devices.

- *Angle finder* – Roofs are normally built with a specific slant or pitch that helps the water move toward a desired discharge point. When evaluating a roof as a potential solar panel installation point, the slope or pitch of the roof must be known. *Figure 2* shows a typical angle finder that can be used to determine the pitch of a roof.
- *Long lengths of strong line (string)* – Line can be used to mark off a potential installation site. Chalk lines can also work.
- *Flashlight* – Useful when examining attic areas.
- *Levels* – Spirit levels are used for traditional leveling checks. *Figure 3* shows a two-foot level and a torpedo level. Laser leveling devices may also be used.
- *Directional tools* – A directional compass can be used to locate true north and true south. If an accurate global positioning system (GPS) is available, it may also be used.
- *Shade analyzer* – These instruments are used to analyze the shade, or potential shade, around

an installation location. One of the more popular devices is the Solar Pathfinder™, which is set up on the proposed site and aligned to true south for any site in the northern hemisphere. (It would be aligned to true north for use in the southern hemisphere.) This instrument reveals any shading at the site at the time of the test. *Figure 4* shows a Solar Pathfinder™.

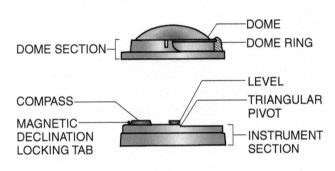

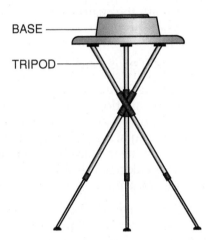

102F02.EPS

Figure 2 Angle finder.

FRAMING LEVEL

TORPEDO LEVEL

102F03.EPS

Figure 3 Short levels.

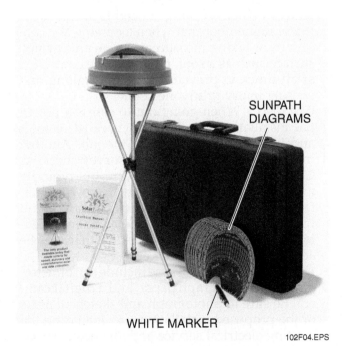

SUNPATH DIAGRAMS

WHITE MARKER

102F04.EPS

Figure 4 Solar Pathfinder™.

NCCER — *Contren® Learning Series* 57102-11

- *Electrical test equipment* – A Category III/IV multimeter may be required for voltage measurements.
- *Ladders* – A conventional 20-foot extension ladder is often all that is needed to reach the roof of most single-story buildings. *Figure 5* shows a ladder used to gain access to a roof area.

> **WARNING!**
>
> Workers climbing onto a roof or up the side of a building must wear the approved fall protection equipment. The fall protection equipment must be anchored to an approved point above the worker. Make sure that the duty rating on any ladder used will be sufficient to support the worker.

- *Aerial lifts* – A mechanical lift (*Figure 6*) may be required to safely reach the roof of a two- or three-story structure without internal access. Some high buildings with flat roofs have a stairway and a hatch or doorway for access onto the roof.

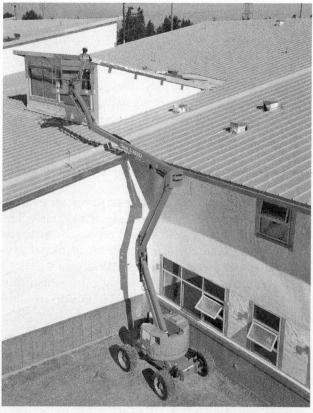

102F06.EPS

Figure 6 Aerial lift being used outside a structure.

> **WARNING!**
>
> Check local safety guidelines about wearing fall protection equipment when working from an aerial lift.

2.0.0 DETERMINING CUSTOMER NEEDS

When assessing the solar power needs of a customer, one of the first things to consider is the customer's budget. After the financial limits have been established, the next step is to examine the customer's energy loads and find out what type of system they are interested in. Will this be a stand-alone system such as for a hunting camp? If it will be tied into the local utility, will it be a utility-dependent grid-tied system or will it be supplied with battery backup as a grid-interactive system? Document the following information:

- The amount of power currently used
- The amount of power expected from the new solar panel system
- The amount of power any essential loads must receive from another source, such as a generator or battery system
- Whether or not a battery bank will be required

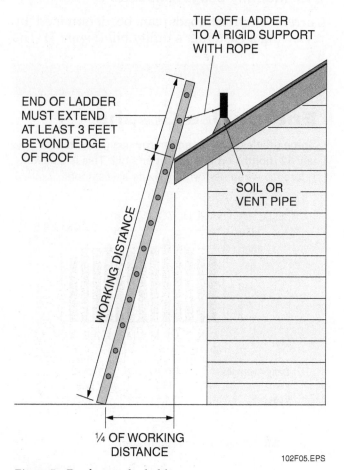

TIE OFF LADDER TO A RIGID SUPPORT WITH ROPE

END OF LADDER MUST EXTEND AT LEAST 3 FEET BEYOND EDGE OF ROOF

SOIL OR VENT PIPE

WORKING DISTANCE

¼ OF WORKING DISTANCE

102F05.EPS

Figure 5 Roof access by ladder.

2.1.0 Energy Loads

Performing a site assessment requires an evaluation of the customer's current electrical loads. The installation of a PV system will reduce these loads and their associated utility costs. Additional savings can be realized by suggesting other energy conservation strategies during the customer interview. The following is a list of energy-related questions to consider:

- When is the building occupied and by how many people? Discuss simple strategies for reducing energy use both when the building is occupied and when it is unoccupied. These include installing programmable thermostats, replacing incandescent bulbs, and turning off the lights when leaving a room.
- Is the hot water heater electric? Point out that insulating a hot water tank can save considerable energy. A solar thermal system can also be considered.
- If meals are prepared at the address, when are they prepared? How are the dishes washed? Electric stoves and dishwashers use large amounts of electricity. Point out the energy savings offered by gas or propane appliances.
- How many loads of laundry are done in a week, and when are the loads run? Electric dryers also use large amounts of electricity. The customer might consider the use of drying racks or a clothesline to conserve energy.
- How old are the appliances? Older appliances are normally much less efficient than current models. Recommend upgrading to energy-efficient equipment whenever possible.
- How many computers, televisions, and entertainment systems are being used, and when are they usually turned off? Discuss the energy wasted by devices such as chargers, televisions, and stereo systems. Point out that up to six percent of the energy used by a residence can be saved by simply unplugging these loads.
- What kinds of heating and cooling systems are used? If the systems are electrical, how old are they, and what is their current condition? Point out the advantages of upgrading to energy-efficient equipment.

- How many ceiling fans are used on a regular basis? The use of ceiling fans can reduce the need for air conditioning.
- How old is the structure, and what is its condition? How well is it insulated? Note that the installation of additional insulation and weatherstripping can offer significant energy savings.
- Are there any plans to replace the roof in the near future? If there are plans to replace the roof, it is best that this be done before the installation of the PV array.
- What kind of seasonal changes can be expected at the location? Weather plays a major role in a customer's electrical usage.
- Is there any equipment that must receive electricity all the time? Essential equipment may include heating or air conditioning equipment, lighting, refrigeration equipment, and security systems.

Many utilities offer free energy audits that help to identify energy waste in a home and reduce the overall energy use. Energy audits analyze the heat loss (or gain) in a home and provide recommendations for additional weathersealing, insulation, and energy-efficient equipment.

2.1.1 Monthly Usage Amounts

Current electrical loads can be determined by reviewing a customer's utility bill (*Figure 7*). The

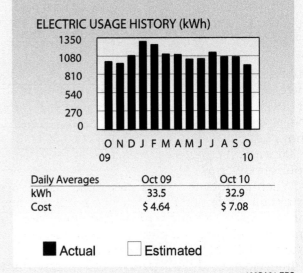

On Site

Energy Use Graphs

Some utilities provide an energy use graph for the last 12 months on the customer's bill. This is a useful reference when performing a site assessment.

ELECTRIC USAGE HISTORY (kWh)

Daily Averages	Oct 09	Oct 10
kWh	33.5	32.9
Cost	$ 4.64	$ 7.08

■ Actual ☐ Estimated

102SA01.EPS

ROY K SMITH
238 WOODVILLE RD

Account Number	000225XXXX
Verification Code	6
Bill Date	02/26/2010
Current Charges Past Due After	03/23/2010

Service From: FEB 04 to FEB 25 (21 Days)   Your next scheduled meter reading will occur between   MAR 30 and APR 05

PREVIOUS BILL AMOUNT	PAYMENTS (-)	NEW CHARGES (+)	ADJUSTMENTS (+ OR -)	AMOUNT DUE (=)
$137.71	$137.71	$109.56	$0.00	$109.56

METER NUMBER	METER READINGS: PREVIOUS	PRESENT	MULTI-PLIER	TOTAL USAGE	RATE SCHEDULE DESCRIPTION	AMOUNT
115424	39695	41079	1	1,384 KWH	RE - Residential Srv,All Electric	109.56
					Amount Due	109.56

Electricity Usage	This Month	Last Year
Total KWH	1,384	1,597
Days	21	29
AVG KWH per Day	66	55
AVG Cost per Day	$5.22	$4.06

Our records indicate your telephone number is 123-456-7890 . If this is incorrect, please follow the instructions on the back of the bill.

A late payment charge of 1.5 % will be added to any past due utility balance not paid within 25 days of the bill date.

For Correspondence: PO BOX 1090 CHARLOTTE NC 28201-1090

102F07.EPS

Figure 7   A monthly electric billing statement for a residential service.

billing statement shows the past and present meter readings. The bill also shows the amount of electricity used over the past month and the amount of electricity used daily. Usage is shown in kilowatt-hours (kWh). One kW equals 1,000 watts (W). One kWh equals 1,000 watts per hour of power used.

When performing a site assessment, it is best to obtain copies of the customer's electric bills for the past 12 months. Most customers can log onto their electric company's website and download this information on demand.

Interviewing the customer is also important. The power bill shown in *Figure 7* is for a ranch-style

house built in 1989. After examining the utility bills, it was found that the residence used an average of 1,632kWh per month over the past two years. The daily average was around 54kWh. The power company classifies this home as an all-electric residential service. It has 1,500 square feet of heated living space. The windows have double-pane glass set in wooden frames. The ridgeline of the house runs east and west. There is a two-car garage attached to the west side of the house. The back of the house faces due south. The southern-facing roof (*Figure 8*) gets maximum sun year round. The north side of the house stays shady most of the time. The attic has moderate

 Site Assessment

Figure 8   Roof on the southern side of the house.

insulation. Ridge vents were added when a new roof was installed in 2009.

Only two people live at the residence. One works and is away from home 10 to 11 hours per day. They prefer to keep the thermostat at 72°F or below in the winter and around 84°F in the summer. The heat pump system was replaced in late 2007. Gas logs are used at times during the winter months. There are no plans to make any additions, but there is a plan to install new insulated windows over the next five years. The residence has two televisions, one computer system, two radios, one microwave, one 27-cubic-foot refrigerator, and one 25-cubic-foot freezer. On most days, the evening meal is the only time the kitchen stove is used.

According to the local weather data, the house is located in a region where the average temperature is fairly mild. The temperatures range from the high 30s in January to the high 80s in the middle of July. Some winter days may have temperatures in the single digits, but those are rare. Summer days occasionally have temperatures as high as 100°F, but those too are rare. The humidity in this region ranges from around 80 percent in the winter months to around 95 percent in August. Air conditioners are used approximately 65 percent of the year. In the winter, this region gets some snow and ice, but not very often. Snowfalls rarely exceed five inches. Normal winds range from 8 to 10 miles per hour (mph). They may occasionally gust to 30 mph or more.

2.1.2 Amount of Solar Power Wanted by the Customer

The customer must decide how much grid power will be replaced by the PV system. In some cases, this is limited by the customer budget and the available space for the array.

The house in the example uses an average of 1,632 kWh per month. It is located in the 29669 zip code. The central heating and air conditioning system and the hot water heater are the greatest users of power. The customer wants a solar panel system that would cut the monthly amount from the grid by 50 percent. Numerous online solar calculators are available to quickly convert this information into a general PV system size. When the customer's zip code, power use, and desired offset are fed into an online solar power calculator (*Figure 9*), the results indicate that the house would need a solar system sized at 6.55kW or 6,550W. Remember that crystalline silicon panels generate about 10W per square foot of space. Therefore, about 655 square feet of roof (6,550W ÷ 10W/sq ft) will be needed to install enough panels to create the desired level of power.

2.1.3 Peak Usage Amounts

Electric companies monitor the power loads as they rise and fall during given periods of time. Hotter weather tends to cause more of a spike in usage than colder weather. This is especially true in the South where winters are shorter and

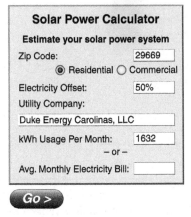

SCREEN 1

Solar Power Calculator

SYSTEM SPECIFICATIONS	
SYSTEM SPECIFICATIONS FOR: PELZER, SC	
UTILITY: DUKE ENERGY CAROLINAS, LLC	
SOLAR IRRADIANCE:	5.17 KWH/SQ METER/DAY
AVERAGE MONTHLY USAGE:	1.632 KWH/MONTH
SYSTEM SIZE:	6.55 KW
ROOF SIZE	654 SQ FT

SCREEN 2

102F09.EPS

Figure 9   Online solar power calculator.

NCCER — *Contren® Learning Series*   57102-11

summers are both longer and more humid. Regardless of location, the electrical loads spike in hot summer weather as customers use more air conditioning.

The rates a power-generating company charges are regulated by the states in which they operate, but the rates can still vary to some degree. The cost of fuels used in the power plant can also affect the rates a power company charges for its electricity. In addition, the rate charged for one kWh may increase when electricity is sold from one company to another.

When determining the electrical needs of a customer, one of the things to look at is how the customer's usage aligns with the power company's peak load charts. The customer's usage is often high when the power company's load is also high, such as late in the afternoon on a hot summer day when people are coming home from work and turning on the air conditioning. This is an ideal time for the customer to use electricity provided by an alternate power source, such as solar panels. As a result, the customer saves money and the power company's load is reduced during a peak period.

2.1.4 Essential Loads

Essential loads often include lighting, refrigerators/freezers, and heating and air conditioning systems. Security or computer systems may also be essential loads. Essential loads require generator or battery backup. The customer must decide how long these loads would need backup power if power from the grid is lost.

After the essential loads have been identified, the amount of electrical power they need must be calculated. Once the electrical needs are known, determine if storage batteries can be used. If batteries can be used, a determination must be made about how much capacity will be needed, when it will be needed, and how long it will be needed. Solar panel systems with storage batteries are rated on their autonomy. Autonomy is defined as the length of time the system can operate on battery power alone. A battery system with two days of autonomy can support the loads for two days without dropping below the allowable depth of discharge. More days of autonomy will require more storage batteries, which in turn will require a larger storage area. More batteries also mean higher project costs.

2.2.0 Available Sunlight

After assessing a customer's electrical needs, the next thing to determine is whether or not there is a location on the property that will provide enough sunlight. In order to operate properly, the panels require about six hours of intense sunlight. The hours of sunlight early in the morning and late in the afternoon have little effect on the panels, but the time of year does have an impact. Summer sunlight is much more intense than winter sunlight. *Figure 10* shows the effects of seasonal changes on the sunlight falling on a house.

For locations north of the equator, orientation on the Earth is referenced to the North Pole. Compass angles are referenced from either magnetic north or true north. The difference between the two depends upon the distance from the equator. The closer it is to the true north, the less difference there is between the two angles.

When considering the location of the sun in reference to a structure north of the equator, stand facing south. A handheld compass can be used to show true south. When facing true south, the sun rises on the left and sets on the right. Solar panels are most effective when they are facing true south and are not shaded. When assessing a potential site, look for an installation location where the panels can receive direct sunlight. Check for any buildings or trees that could shade the panels during the midday hours. Consider the future growth of nearby trees and shrubbery. Ask the customer if there is any anticipated construction near the site that may result in shading.

2.3.0 Mounting Options

A number of options are available for mounting solar panel arrays. Some are mounted on stand-alone poles out in a field (*Figure 11*). Others are mounted on frames or racks that support them at different angles. Some racks are motorized, which allows them to move the panels in order to follow the sun as it crosses the site.

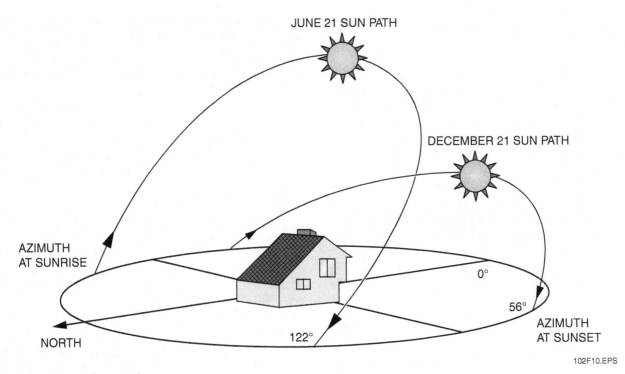

JUNE 21 SUN PATH

DECEMBER 21 SUN PATH

AZIMUTH
AT SUNRISE

0°

56°

AZIMUTH
AT SUNSET

NORTH

122°

102F10.EPS

Figure 10 Seasonal sun paths.

102F11.EPS

Figure 11 Pole-mounted panels.

Any array of solar panels mounted on the ground must be protected from vandalism and damage from natural causes. One way to protect the solar panels is to install them at least eight feet above ground level. If fencing is used to protect ground-mounted panels, the fencing must be far enough away from the panels to prevent shading. Fencing must also be grounded. All power cables between the array and the building must be protected in conduit.

Many solar panel arrays are mounted on the roof of a structure. Some solar panels are integrally-mounted, which means they are built into the roof itself. Other panels are standoff-mounted, which allows for better airflow by providing a gap between the panels and the roof surface. Standoff-mounted panels are typically installed three to five inches above the roof surface. This keeps the panels cooler and results in higher efficiency. *Figure 12* shows homes with integrally-mounted and standoff-mounted solar panels.

Solar panels are currently being used on residential carports and on carport covers in large parking lots (*Figure 13*). Large parking lots are typically unshaded, which makes them ideal locations for solar panels. This form of installation places the panels well above the ground, wastes very little space, and protects vehicles from the damaging effects of sunlight.

INTEGRALLY-MOUNTED PANELS

STANDOFF-MOUNTED PANELS

102F12.EPS

Figure 12 Integrally-mounted and standoff-mounted panels.

102F13.EPS

Figure 13 Solar panel carport.

3.0.0 MOUNTING SURFACE INFORMATION

The mounting surface information is some of the most important data collected during a site assessment. When documenting this data, be as specific as possible. Take pictures of each item on the checklist. This will make them easier to find when laying them out in the final report. At a minimum, the checklist data should include the following information:

- Document the type, age, and condition of the roofing material or mounting surface. Evaluate the strength of the supporting structures and describe any needed reinforcements.
- Document the slope of the roof or mounting surface(s).
- Note the accessibility to the proposed array location. Keep in mind that machinery may need to be brought in to access the roof.
- Document the location of overhead and underground power lines, fencing, close passageways, and flower beds or other obstacles in relation to the installation site.
- Record all potential shading information. A Solar Pathfinder™ or similar instrument can be used to gather shading data. A sketch of the facility is usually required. Trees or buildings that may shade the location should be sketched and/or photographed.
- Measure the proposed installation site and draw or photograph all obstructions.
- If the panels cannot be mounted side by side to form one large array, they may need to be separated and mounted in smaller arrays that will supply the required voltage when their outputs are combined. The square footage of these smaller arrays is measured and then added. If the arrays will be mounted on the ground, the same process applies. There will be a place on the site survey checklist to list the estimated square footage required. *Figure 14* shows typical measurement references on a residential structure.
- Note the azimuth direction of the proposed installation. Keep in mind that if the site is in the northern hemisphere, the sun will be to the south. Document which way the ridgeline of the structure is aligned (east to west, north to south, northeast to southwest, etc.).

3.1.0 Inspecting Proposed Installation Areas

Any area where solar panels will be installed must be able to support the weight of the new equipment. It must also serve as a solid anchor for the solar panels. Remember that solar panels and the surfaces on which they are mounted are exposed to various stresses such as expansion and contraction, wind loads, and drag. They must also be mounted at angles that will expose them to the most direct rays of the sun.

70 FT

22 FT

15 FT

6" BATHROOM
VENT
12" FROM
EAVES

AVAILABLE SITE FOR ARRAY ON SOUTH-FACING ROOF
(20 FT × 50 FT)

102F14.EPS

Figure 14   Measurement points on a residential structure.

Building codes require that roofs be designed to support specific load amounts. Anything that is mounted on the roof increases the load. Structures built in colder climates must have roofs that will support large amounts of snow and ice. The roofs on those structures have much steeper slopes. The slope or pitch of a roof can be checked with a pitch and angle gauge (*Figure 15*).

PV support structures must be strong enough to support and anchor the new equipment. Pay close attention to the condition of the roofing materials. *Figure 16* shows the load areas of a given residential structure. The materials used in the structure must meet certain load limits. A dead load is the weight of permanent, stationary construction and equipment included in a building; in other words, the building materials themselves. A live load is the total of all moving and variable loads that may be placed on a building. Adding an array of solar panels to a roof means that the weight of the materials and the weight of the workers will be on the roof during the construction. Verify that the roof will support the additional loads created by the solar array.

When inspecting the structural condition of the roofing materials, take a close look at the condition of the materials themselves. Visual inspection is the primary method of evaluation. If the rafters and roof sheeting are made of metal, check for rust, leakage stains, and loose connectors. If the roofing materials are made of wood, look for discolored areas, cracks, and water stains. Stick a knife blade into any suspected area to see if it is soft. Even 20-year-old wood should not be soft. Look for any visual signs of mildew and note any smell. If the installation site is a roof that is due to be replaced within the next 10 years, the owner should consider replacing it before the solar panels are installed. Most solar panels are warranted for 20 years. If the roof needs to be replaced during that period, the panels would have to be removed and then reinstalled.

As previously noted, the roofing materials must support the weight of the solar panels and their associated devices. When solar panels are added onto any surface area, more weight (loading) is placed on the installation surface. That weight includes the weight of the panels themselves and the hardware used to support and anchor them.

Solar panels are made of various materials. Each manufacturer tries to build the most lightweight panel that produces the greatest amount of electricity. Solar panels are categorized by the amount of power (voltage or wattage) they can produce. The following is a list of facts about a typical 240W solar panel:

- *Length* – 63.1" (1,602 mm)
- *Width* – 41.8" (1,061 mm)

SLOPE CHECK

PITCH CHECK

102F15.EPS

Figure 15 Checking the slope and pitch of a roof.

- *Depth* – 1.57" (40 mm)
- *Weight* – 44.1 lbs (20 kg)

Solar panels producing less power will be lighter and smaller. Many of them produce only 6V or 12V of direct current (DC).

3.2.0 Inspecting Attic Spaces

Access to the underside of the roof of a residential structure is usually gained through a pull-down ladder inside the building. Always make sure to thoroughly inspect all ladders before climbing them. Some attic spaces have flooring, but most do not. Some have limited flooring used for additional storage. If the attic is not floored in the area under the proposed installation site,

be very careful when moving around in the attic. Only walk on top of the ceiling joists. Boards or plywood cut to one-foot widths can be brought up into the attic to serve as temporary flooring. Lay those boards across the ceiling joists to provide reasonable standing or sitting space while the underside of the roof is being evaluated. *Figure 17* shows the underside of a typical residential roof, as seen from the attic.

> **WARNING!**
> Attic insulation can create breathing problems. Attic areas may also be dusty. Wear a respirator as needed. Insulation also irritates exposed skin. Wear long pants, long-sleeved shirts, and gloves in attics. Carry a good flashlight for additional lighting.

Gaining access to the underside of buildings with high ceilings presents a different set of issues. In many cases, the bottoms of the roof trusses can be 12 to 30 feet or more above the floor or ground. Extension ladders may be safely used on the lower trusses. For the higher trusses, some type of aerial lift (*Figure 18*) is needed. A customer running a business from a structure with high trusses will often have an aerial lift that is used for routine overhead maintenance. For liability reasons, the customer's operator may be needed to operate the lift.

> **WARNING!**
> Check local safety guidelines about wearing fall protection equipment while working from an aerial lift.

3.3.0 Roofing Materials

When assessing an installation site that will have the panels installed on the roof of a structure, check the roofing material(s) used on the structure. Standard shingles are used on many homes and businesses, but some structures have metal roofs. Structures in the southwest often have ceramic (Spanish) tiled roofs. Installation on a tile roof requires special precautions to avoid damaging the tile and is typically performed by a specialized installer. Some flat roofs are covered with sheets of underlayment materials that are then covered with heated tar and some type of

GIRDER LOADS (POUNDS PER SQUARE FOOT)

	LIVE LOAD*	DEAD LOAD
ROOF	20	10
ATTIC FLOOR	20	20 (FLOORED)
		10 (NOT FLOORED)
SECOND FLOOR	40	20
PARTITIONS		20
FIRST FLOOR	40	20 (CEILING PLASTERED)
		10 (CEILING NOT PLASTERED)
PARTITIONS		20
TYPICAL – REFER TO LOCAL REQUIREMENTS		

EXAMPLE:

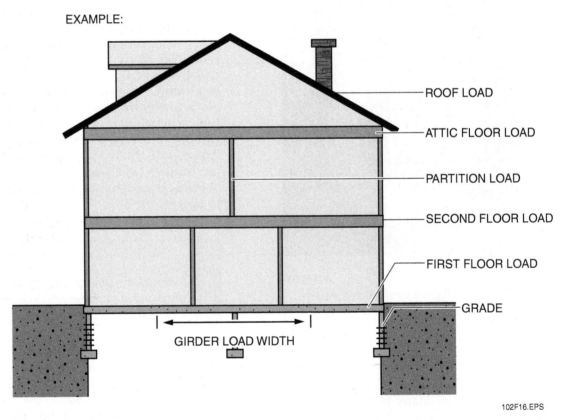

102F16.EPS

Figure 16 Load areas.

protective material, such as pea gravel. *Figure 19* shows different types of roofing materials often used on sloped roofs.

> **CAUTION**
>
> Be extremely careful with all roofing materials. Minor damage can cause major leaks inside the structure.

3.4.0 Spacing

When evaluating a site for solar panel installation, the space needed to mount the array depends on several factors, including the size and power output of the panels and the array configuration. For example, one solar panel may have an output of 12 VDC. To increase the level of voltage from an array of solar panels, two or more panels are connected in series. As a result, four panels connected in series should produce close to

NCCER — *Contren® Learning Series* 57102-11

Figure 17 Attic view of the roof's underside.

provides a general idea of the manner in which the panels would be laid out and connected.

A typical 12V solar panel measures roughly 60 inches in length and 26 inches in width. In the example shown, the four panels in each row would extend at least 240 inches horizontally. When measured vertically, the four rows would measure at least 104 inches. Since there may be a couple of inches of space between the panels, round the horizontal distance up to 246 inches. Round the vertical distance up to 110 inches. Multiplying the horizontal measurement (246 inches) by the vertical measurement (110 inches) produces a space of 27,060 square inches. Since one square foot is equal to 144 square inches, divide 27,060 square inches by 144 to get the overall footprint of 187.92 square feet.

Because of the arrangement of the panels shown in *Figure 20* (laid out end-to-end and stacked in four rows), the space will have to be roughly twice as long horizontally as it is vertically. If the panels cannot be laid out as shown, they can be rearranged into other patterns. In some situations, an array of solar panels is placed on one section of roof, and another array is placed on a nearby section. All the panels of a solar system do not have to be placed in the same location. As long as the installed panels receive the desired amount of sunlight, the rows of panels can be separated as needed.

Solar panel dimensions are usually given in metric units. Some panel makers list the dimensions of their panels in both metric and standard units. The easiest way to convert millimeters (mm) into inches is to remember that 25.4 millimeters equals 1 inch. For a typical solar panel that is 60 inches long, its metric length measurement would be 1,524 millimeters. A panel width of 26 inches would convert to a metric width measurement of 660.4 millimeters.

As a rule, a crystalline silicone module (panel) with 10 percent efficiency generates about 10 watts per square foot (or 100 watts per square meter) of illuminated area on the panel. Multiplying the available and usable area of a proposed site by 10 will provide an estimate of the number of panels the area will hold. Measure the area's length and width to get the square footage, and then reduce that figure by perhaps 30 percent to allow for unusable space. For example, if a roof area measures 14' by 25', the square footage would equal 350 square feet. Multiply that amount by 70 percent to arrive at 245 square feet of usable space. Round that number up to 250 square feet and multiply it by 10 watts per square foot. The roof can support panels capable of generating about 2.5 kWh of power. If more power is needed

Figure 18 Aerial lift being used inside a warehouse.

48 VDC. To increase the level of current from an array of solar panels, several strings of panels are wired in parallel. The result is an array that eventually creates a footprint of a certain size. *Figure 20*

SHINGLE ROOFING

MACHINE-CRIMPED PANELS

SNAP-TYPE SEAL

METAL ROOFING

CERAMIC SPANISH TILES

SIMULATED WOOD SHINGLE ROOF

TILE ROOFING

102F19.EPS

Figure 19 Shingle, metal, and tile roofs.

from an array, more space will be required or more efficient panels must be used. In large arrays, a walkway space must be left between panels so that maintenance personnel can move between them.

Tilted panels have additional spacing requirements to avoid shading. Be sure that the installation area allows for adequate spacing when panels require tilting. See *Figure 21*.

When the panels must be tilted to a specific angle to maximize sun exposure, the spacing is calculated by dividing the height of the tallest point on an array by the angle of the sun (a) to the horizontal

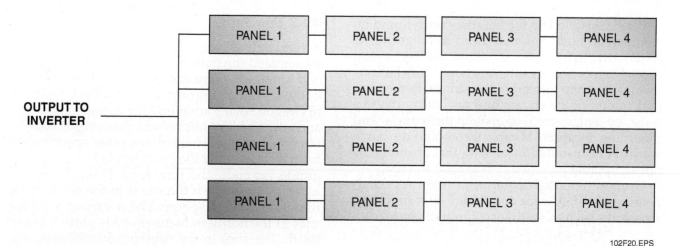

Figure 20 Typical solar panel array arrangement.

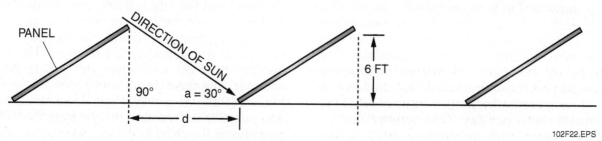

Figure 21 Tilted panels require additional spacing to avoid shading.

surface under the arrays (*Figure 22*). To find the distance (d) back to the base of the next array, divide the height of the array in front of the angle by the tangent of the angle. For example, assume that the direction of the sun creates an angle of 30 degrees. The tangent of 30 degrees is 0.57735. Next, assume

that the highest point on the front array is 6 feet. The offset distance is calculated as follows:

Distance = height ÷ tangent of angle a

Distance = 6 feet ÷ 0.57735
= 10.4 feet or approximately
10 feet, 5 inches

3.5.0 Anchoring

A panel with a surface area of 10 square feet can produce an uplift load of 500 pounds of force pulling against its mounting hardware. Installation sites, such as the roof of a structure, must be able to withstand that kind of force. The type of mounting hardware and the number of mounting points depend on the solar panels being used. The more mounting points used under a panel, the less force is exerted on the roof. However, excess mounting points should be avoided as each mounting location represents a roof penetration that increases the chance of leaks. Makers of mounting hardware provide engineering data with the hardware. If there are any concerns about the strength of the structure at a proposed site, call in a structural engineer to make the final decision. Some local building inspectors may require proof of engineering assistance before they sign off on an installation.

Figure 22 Calculating panel spacing for a tilted array.

Panels installed on the flat roofs often found on commercial buildings may need to be anchored differently than those mounted on residential roofs. Such panels are often anchored with ballasts (heavy weights) instead of lag screws. Remember that mounting hardware and anchoring devices must be engineered to match the panels and mounting location. Mixing the mounting hardware of one installation job with that of another is not acceptable.

Ground-mounted equipment requires engineering assistance for the design of supporting structures such as concrete columns or piers.

4.0.0 ACQUIRING AND INTERPRETING SITE SOLAR DATA

Solar panel installation areas need to receive an average of six full hours of sunlight for best system performance. The best times are between 9:00 AM and 3:00 PM. To acquire solar data about a site, its exact position must be known. After that information is known, the sun's path over the location can be plotted. Any shaded areas must be documented.

4.1.0 Identifying Sun Paths and Intensity Levels

There are numerous websites offering a variety of solar-related information and calculators. The zip code entered earlier into the solar power calculator was used to determine the rough location of the structure. If latitude and longitude data is needed, there are websites that will provide that information when a complete address is entered. With exact latitude and longitude data in hand, overhead aerial views of a site can also be obtained from several online mapping sites. All such information must be obtained before going onto a site.

4.1.1 Sun Intensity

The level of sunlight intensity at a site can also be obtained online when exact position data is provided. Sunlight intensity is called irradiance and is measured in watts per square meter. One square meter equals 10.7639 square feet. When doing a site assessment, the intensity of the sun at the site location must be determined. The solar power calculator for the example home discussed previously indicated that the location should have a solar irradiance value of 5.17 kWh per square meter per day. That number changes slightly from month to month as the seasons change.

4.1.2 Sun Paths

Several online programs can be used to produce a customized sun path chart by entering the latitude and longitude numbers for a given location. One of those online programs is run by the University of Oregon Solar Radiation Monitoring Laboratory and is available at http://solardat.uoregon.edu. To obtain a sun path chart or any other specific data from this site, enter the location's latitude and longitude, zip code, and time zone. This information tells the computer if the site is in the northern or the southern hemisphere. The reference point for sites in the northern hemisphere is north looking south. For sites in the southern hemisphere, the reference point is south looking north. For solar panels to be effective, they must be facing in the direction that provides the most intense sunlight. For this module, the site will be north of the equator, and the panels will need to be facing south.

4.1.3 Elevation Views of Sun Paths

When specific site data is entered into the University of Oregon's online program, a screen appears to select different charts related to the available solar data for the location. *Figure 23* shows an elevation view of the sun's intensity over the example home between December and June. This chart was created specifically for the example home shown and discussed earlier in this module. A similar chart can be obtained for the months between June and December.

Note that this is an elevation, or altitude angle, view of the sun crossing a site. The view is from the north looking south. The sun's paths traced on this image are laid over a grid. The horizontal grid lines are the solar elevation angles (shown along the left side of the chart). The vertical grid lines are the azimuth angles around the site (shown at the bottom of the chart). The sun rises on the left (east) and sets on the right (west). The example site is at the 180-degree mark at the bottom center of the chart.

Figure 23 shows that the sun begins to rise at a different point on the December 21 path than it does on the June 21 path. The sunsets are also different. At any sunrise, the elevation angle of the sun from the site is 0 degrees. When the sun is directly over the site, the elevation angle is at its highest point. At sunset, the elevation angle is back to 0 degrees or 360 degrees. The site is getting the most sun when the sun is directly above it. The most direct sunlight is between the hours of 9:00 AM and 3:00 PM. Use a straightedge and pencil to draw a straight line from the 9:00 AM position on the chart to the site. Do the same from the 3:00 PM position to the site. With those lines

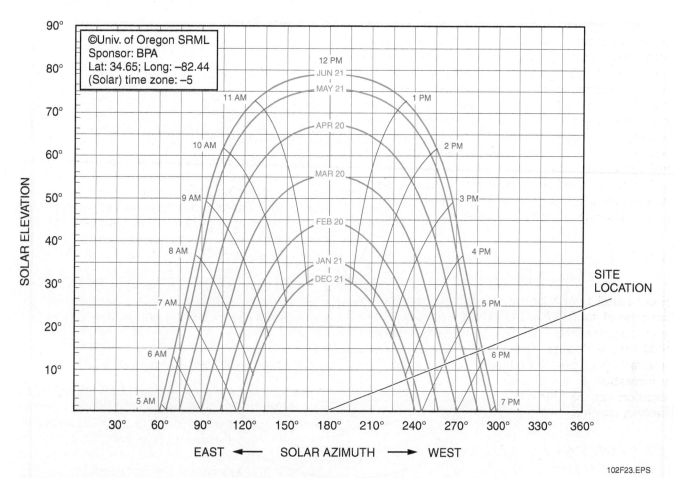

Figure 23 Elevation view of a sun path chart for a specific site.

drawn, it is clear that the site gets much more sunlight during the summer months than it does during the winter months.

Look at the 30-degree mark on the left, and then follow the horizontal line to the right. The sun does not get much higher than about 32 degrees on December 21 but reaches an elevation angle of almost 80 degrees at noon on June 21. The sun's path can be plotted at any given point in time on this chart. For example, determine the sun's elevation angle at 2:00 PM on April 20. The sun is at an elevation angle of 55 degrees. Plotting the sun's path at different times of year and at different times of day is necessary in order to determine where to install the panels. In some cases, shading may surround a site, but if the sun's position is already above the shade, the shading will not interfere with the proper operation of the solar panels.

4.1.4 Overhead Views of Sun Paths

Another chart available from the University of Oregon's website shows an overhead view of the sun's movements across the targeted site (*Figure 24*). The months are shown by the horizontal arcs that move across the circle from left to right. The hours of the day are shown by the vertical rays that cross the months. The days are longer in the summer months and shorter in the winter months.

4.2.0 Shading at the Customer's Location

A shade analysis instrument, such as a Solar Pathfinder™, is used to determine whether shade covers any part of the selected site during the hours of 9:00 AM and 3:00 PM. *Figure 25* shows a Solar Pathfinder™ set up on a proposed installation site.

After the base of the Solar Pathfinder™ has been set up and leveled, one of the supplied charts is selected to use in plotting the sun's path and any related shaded areas. The Solar Pathfinder™ kit comes with sun path charts for surfaces with different degrees of slope. It also includes charts for different latitudes. When a Solar Pathfinder™ is set up, the glass dome is removed, and the selected sun path chart (*Figure 26*) is installed onto the base.

Notice that the horizontal and vertical grid lines on this chart are very similar to those shown on the sun path chart in *Figure 24*. The horizontal lines represent the months. The vertical lines signify the hours of a solar day.

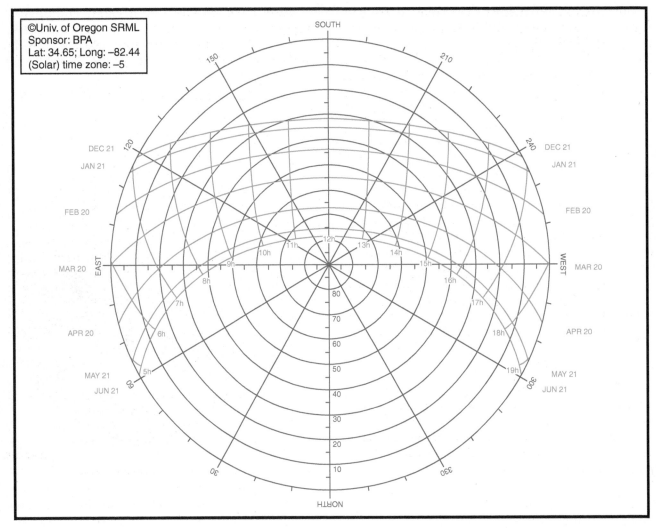

102F24.EPS

Figure 24 Overhead view of a sun path chart for a specific site.

REFLECTIVE DOME

SOLAR PATHFINDER™
SHADE ANALYZER BASE

ADJUSTABLE LEGS

102F25.EPS

Figure 25 Solar Pathfinder™ in use.

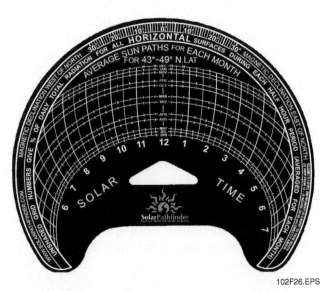

102F26.EPS

Figure 26 Solar Pathfinder™ chart for 43°–49° North Latitude.

Compasses align themselves to magnetic north. If the site being evaluated is at a location where true north and magnetic north are not the same, the Solar Pathfinder™ will have to be aligned to compensate for the difference. The upper edge of the Solar Pathfinder™ chart shows the degrees of magnetic declination. The Solar Pathfinder™ base can be unlocked and moved to align the chart with whatever amount of magnetic declination is needed for the site. After the realignment has been completed, the base is again locked down. *Figure 27* shows a sun path chart on the base of a Solar Pathfinder™. The chart has been adjusted for 11° east. Magnetic declination values can be found by entering the site zip code in the magnetic declination calculator available on the National Geophysical Data Center (NGDC) website at **www.ngdc.noaa.gov**. For example, the sample house discussed previously is located at zip code 29669. According to the NGDC, it has a magnetic declination factor of approximately 6° west.

After the chart has been installed and aligned, replace the dome. *Figure 28* shows an overhead view of a chart under the dome. Notice how the reflections of the trees surrounding the dome can

MAGNETIC DECLINATION ADJUSTED FOR 11° EAST

102F27.EPS

Figure 27 Sun path chart installed on a Solar Pathfinder™ base.

Did You Know?

A magnetic compass may not provide a true reading when used on a metal roof. Use a digital compass. In addition, an instrument tripod may slide on a metal roof. Use blocking and/or wedges to level and secure the instrument.

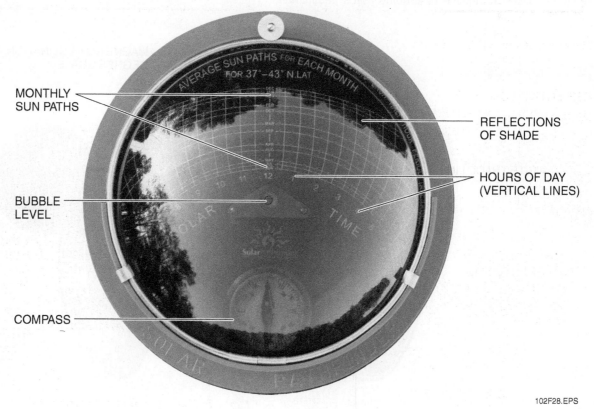

102F28.EPS

Figure 28 Sun path chart under a Solar Pathfinder™ dome.

be seen over the grid lines of the installed chart. The bubble level near the center of the instrument is used for leveling it after it has been set. The compass at the bottom is used to orient the instrument.

A white marker is inserted through the open spaces surrounding the dome's bottom and used to trace the outline of the shaded areas onto the chart. Trace lightly to avoid breaking the marker. *Figure 29* shows a closer view of the base and dome of the Solar Pathfinder™.

After the initial trace has been completed, remove the chart and go over the light trace marks again with the white marker to make them easier to see. Be sure to note on the chart whether any of the trees are evergreen or if they lose their leaves in the fall. Also note the date and time that the trace was made. *Figure 30* is a Solar Pathfinder™ chart that was traced to show the shaded areas around a site.

In some cases, the instrument may have to be moved to several locations around the site until the best location is identified. To get the best results from the panels, take readings from all four corner points to ensure that all possible shading will be recorded. If the structure has not yet been built, position the instrument as close as possible to the proposed location.

102F29.EPS

Figure 29 Solar Pathfinder™ base, dome, charts, and markers.

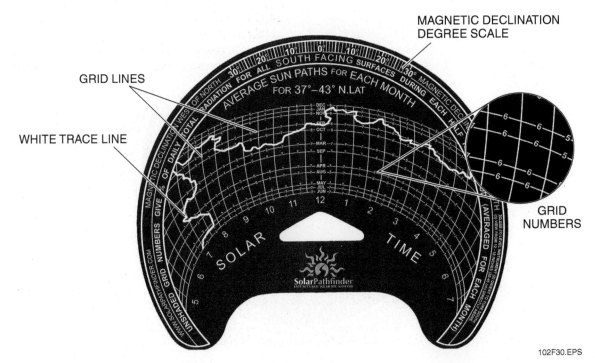

102F30.EPS

Figure 30 One of the Solar Pathfinder™ charts.

After the charts have been plotted, the next step is interpreting the data. The chart indicates how many hours (or half hours) per day of sunlight are available at the site. Note that there are grid numbers on the horizontal month lines going across the chart in *Figure 30*. Those numbers are higher near the 12 o'clock vertical marker. The numbers are lower in the morning hours and in the afternoon hours. Each of these numbers represents a percentage of sunlight at a given hour in a given month, and when they are added up across the arc, they total 100 percent. After the shaded areas are traced out on a chart, add the numbers in the unshaded areas on each month's horizontal arcing line. On the bottom of the chart in *Figure 30*, an arcing line is shown for the month of June. Notice that there is a small amount of shading on the eastern (left) side of the chart, but none on the right side. When all of the unshaded grid numbers are added up, the total comes to 98, which means that during the month of June, this site will receive 98 percent of the available sun crossing it. Now look at the month of October. A considerable amount of shade is covering that arcing line. Any panels installed at this location would see minimal sun during October. During the summer months, the sun's elevation is higher. In the winter, the sun's elevation is lower. Because of this lower elevation in the winter months, any trees near the site will block the sun from the panels. This information must be included on the report given to the PV system designer.

5.0.0 LOCATIONS FOR BOS COMPONENTS

A site assessor must identify suitable locations for both the array and the balance-of-system (BOS) components. Keep in mind that some of the BOS components are designed to be out in the weather, while others are not. Any disconnects must be easily accessible from ground level. To reduce the loss of power over distances, the components should be located close to one another. For small PV systems, the BOS components may be single units that take up minimal space. When larger solar panel systems are installed, multiple BOS components may be required. Racking systems are available for use with larger systems. This module covers the following BOS components:

- Support and security structures
- Combiners
- Charge controllers
- Batteries
- Inverters
- Disconnects
- Grounding

5.1.0 Support and Security Structures

All electrical equipment must be protected against accidental contact by unqualified personnel. Ground-mounted arrays require security fencing. In addition, the array's structure must have a firm foundation. If the array will be motorized in order to track the sun, there must be sufficient space at the site to allow for its movements. When multiple arrays are installed close to each other, they must be separated by enough space to ensure that they will not shade each other. *Figure 31* shows several arrays being erected in a field. The concrete piers used to support the arrays were designed with engineering assistance to ensure that they will safely withstand wind loading.

A security fence is typically installed well away from the arrays and their associated equipment.

102F31.EPS

Figure 31   Arrays in a field.

The fencing and any ground cover near field-mounted arrays cannot be allowed to shade the panels. Anyone performing work inside the fence must be briefed on any potential electrical hazards associated with the equipment. The wiring from the panels is run inside protective conduit that most likely needs to be buried in trenches. Site assessors must identify and mark any underground utilities if the ground must be trenched. All states require that the local utilities be contacted before digging. The "811 Call Before You Dig" program is currently available in about ten states and is expected to become increasingly prevalent.

5.2.0 Combiners

The output of the panel termination boxes may be routed into a circuit combiner. A circuit combiner takes multiple inputs and produces a single output. *Figure 32* shows the terminals inside a typical combiner. This combiner can handle the inputs from up to six strings of panels or inputs from six single panels arranged in parallel. Each input is fused for a maximum of 20A.

The combiner measures 14" high, 12" wide, and 6" deep. A standard fiberglass NEMA 4X enclosure weighs 14 pounds. Metallic combiners weigh two to three pounds more. Combiners with more input terminals are slightly larger. Those made by other companies are of similar size and weight. Combiners are usually mounted near the panels.

They are designed to withstand temperatures ranging from 15°F to 130°F and humidity up to 100 percent.

Some combiners also serve as a disconnect between the panels and the other BOS components. The disconnect switch lever would be on the outside of the combiner enclosure.

5.3.0 Charge Controllers

When storage batteries are used in a solar PV system, the system has one or more charge controllers to regulate the current flow to the batteries.

Different solar panel systems require different types of charge controllers. Charge controllers vary in size (*Figure 33*). Most charge controllers are relatively small (roughly 12" × 24"). They are often located near the batteries. Charge controllers must be mounted in accessible locations as they typically have readouts that must be monitored and controls that are used to set, check, or reset the device.

5.4.0 Batteries

When assessing a site that will be using solar batteries, estimate the number of batteries needed for the job, then determine the amount of space needed to install them. Batteries should be installed in a protected environment where they will be kept at or near room temperature. They also should be kept clean. Because these batteries

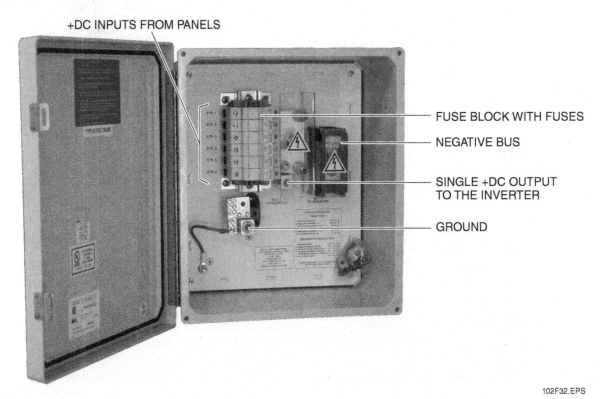

+DC INPUTS FROM PANELS

FUSE BLOCK WITH FUSES

NEGATIVE BUS

SINGLE +DC OUTPUT TO THE INVERTER

GROUND

102F32.EPS

Figure 32   Terminals inside a typical combiner.

NCCER — *Contren® Learning Series* 57102-11

102F33.EPS

Figure 33 Typical charge controller.

typically weigh between 95 and 120 pounds, they require a strong storage structure. Any storage area must be able to withstand the weight of the batteries.

Figure 34 shows a flooded lead acid (FLA) battery and a sealed absorbent glass mat (AGM) battery. The AGM battery is basically maintenance free and does not have special storage requirements beyond installing it in a protected area free from temperature extremes. It is approximately 10 inches long, 7 inches wide, and 13 inches high. Its weight is 91 pounds.

The FLA battery measures roughly 12 inches long, 7 inches wide, and 17 inches high. It weighs about 117 pounds. FLA batteries require venting and refills. Wherever these batteries are installed, enough room must be provided to check and refill them while they are still secured in their mounting racks. The batteries must also have enough space around them for air movement. Batteries are often stored inside a sealed cabinet or room in which the inside air is vented out as heated or cooled air is supplied into the enclosure. The moisture level of the air surrounding the batteries must be kept at an acceptable level, which is usually listed in the vendor manuals. AGM batteries require no maintenance and can be stored in smaller cabinets. *Figure 35* shows a typical cabinet used for storing AGM batteries.

**FLOODED LEAD ACID
(FLA) BATTERY**

**SEALED ABSORBENT GLASS
MAT (AGM) BATTERY**

102F34.EPS

Figure 34 Examples of solar batteries.

5.5.0 Inverters

Figure 36 shows a typical inverter used for residential and smaller commercial applications. The bottom third of this model is used for the incoming and outgoing connections. The upper two-thirds houses the electrical circuits and control devices for converting DC voltage to AC voltage. This inverter weighs about 90 pounds. It is 36 inches tall, 17 inches wide, and slightly less than 10 inches deep. It has a NEMA 3R enclosure. The inverter feeds AC voltage into an AC

Figure 35 AGM battery storage cabinet.

102F35.EPS

POWER STAGE SET(S)

CONNECTION AREA

102F36.EPS

Figure 36 Typical inverter.

disconnect. *Figure 37* shows multiple inverters installed in a special equipment vault for a large commercial PV system.

5.6.0 Disconnects

Each piece of PV equipment requires a switch or breaker to disconnect it from all sources of power. AC disconnects are located near the service panel.

102F37.EPS

Figure 37 Multiple inverters for a large PV system.

DC disconnects are sometimes supplied with the inverter. If they are not, they must be purchased and installed separately. DC circuits require disconnects that are rated to interrupt all current-carrying conductors (both positive and negative).

5.7.0 Grounding

The grounding system is essential to the electrical integrity of a PV array. The roof ground should be a minimum 6 AWG solid copper to provide extra strength. Many ground-mounted PV systems incorporate multiple levels of protection against stray voltages, such as lightning. These are installed in addition to the equipment grounding system, and may include surge protectors and the use of grounded metal fencing for ground-mounted arrays.

6.0.0 DOCUMENTING SITE ASSESSMENT

Most solar installation companies use a checklist to document the assessment data. A typical site survey checklist can be found in the *Appendix*. Checklists should include the following information:

- Contact information
- Electrical needs
- Mounting surface information
- Site sketch

After documenting all the data about where the array panels will be installed, the next step is to document where the balance-of-system components will be placed. This list will be used to estimate the overall cost of the project. Be specific with all measurements. Start at the wires coming off the panels and make a rough determination of the following:

- Where the termination boxes or combiners will be located
- Possible routing of wiring between the array and BOS components
- Where any additional grounding devices can be located
- Where lightning arrestors can be located
- Distance from the termination boxes or combiners to the charge controllers
- Location of batteries (if used), and any ventilation or cooling needed for the batteries
- Distance from the charge controllers to the batteries
- Location of inverters
- Distance from charge controllers to inverters
- Location of DC disconnect switches
- Location of AC disconnect switches
- Any additional electrical generation devices (engine or wind generators)

7.0.0 NEW TECHNOLOGY

When a power plant generates electricity, it is sent out onto what is known as the grid. The electric grid includes all equipment that exists between the generator and the customer. A good percentage of the electricity generated is never used. In 2007, the U.S. Department of Energy (DOE) and other governmental agencies started an initiative to make the national electrical grid more efficient. A smart meter can be installed to notify the grid controller that the customer has little or no need for all the electricity being supplied. One Texas utility company spent $690 million for 3 million smart meters that they expect to have online by 2012. Other utility companies are doing, or planning to do, the same thing. Their objective is to save resources by being more efficient in the way electricity is generated and distributed. *Figure 38* shows a smart meter.

102F38.EPS

Figure 38   Smart meter.

Smart meters can be programmed to shut off electricity to certain appliances within the customer's facility when they are not needed. The end results include a major savings in power production for the utility and power savings for the customer. If the customer has solar power or some other type of alternate power source, the smart meter can be programmed to switch the facility to alternate power when the grid is under a peak demand. *Figure 39* shows a residence equipped with smart meter equipment.

Smart meter technology is already being used in millions of homes across the country. When assessing a structure for the installation of solar panels, determine if it currently has or is expected to use smart meter technology.

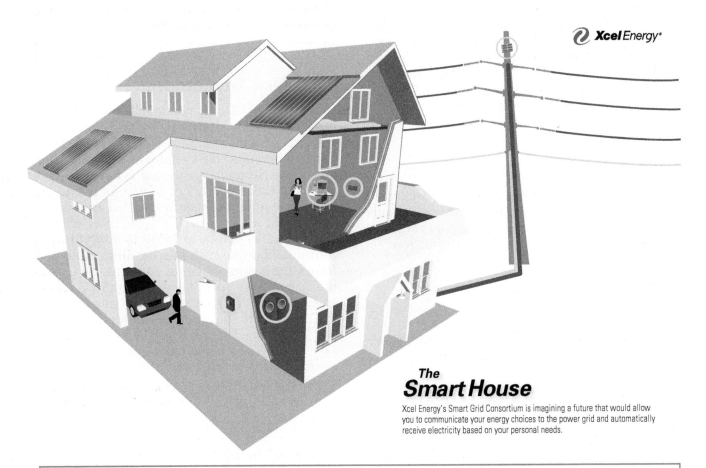

Xcel *Energy*®

The
Smart House

Xcel Energy's Smart Grid Consortium is imagining a future that would allow you to communicate your energy choices to the power grid and automatically receive electricity based on your personal needs.

The potential benefits:
- Lower cost of power
- Cleaner power
- A more efficient and resilient grid
- Improved system reliability
- Increased conversation and energy efficiency

Plug-in Hybrid Electric Car
Xcel Energy is studying how plug-in electric vehicles can store energy, act as backup generators for homes and supplement the grid during peak hours.

Smart Meter
Real-time pricing signals create increased options for consumers.

Smart Appliances
Smart appliances contain on-board intelligence that "talks" to the grid, senses grid conditions and automatically turns devices on and off as needed.

High-Speed Connections
Advanced sensors distributed throughout the grid and a high-speed communications network tie the entire system together.

Customer Choice
Customers may be offered an opportunity to choose the type and amount of energy they'd like to receive with just the click of a mouse on their computer.

100 percent green power? A mix of sources? The cheapest priced source? In Smart Grid City, it could be up to you.

Smart Thermostat
Customers can opt to use a smart thermostat, which can communicate with the grid and adjust device settings to help optimize load management. Other "smart devices" could control your air conditioner or pool pump.

102F39.EPS

Figure 39 Residence with smart meter technology.

SUMMARY

A site assessment is required to determine the customer's needs, identify essential loads, and document the site orientation, shading, and roof condition. The first step in a site assessment is evaluating the customer's budget and the existing electrical loads. This is an ideal time to review other energy-saving opportunities with the customer. Next, the site must be examined to see whether the roof (or other location) is suitable for the installation of an array. The condition of a roof is especially important as a solar array will produce various loads, including the weight of the equipment itself and wind loads.

Site assessment requires the use of a shade analysis device, such as a Solar Pathfinder™. This tool is used to evaluate the shading on the site to select the best location for the array. Solar arrays will not operate properly when shaded. After the sun path charts have been graphed with the shading for the specific site, they are presented to the system designer along with the site survey checklist, sketches of the proposed array location, load information, and potential locations for the BOS components.

Review Questions

1. Fall protection must be worn when working at heights of _____.
   a. 4 feet or more
   b. 6 feet or more
   c. 8 feet or more
   d. 12 feet or more

2. The tool most likely to be used when measuring the length of a very long roof would be a _____.
   a. laser measuring device
   b. carpenter's rule
   c. measuring wheel
   d. Solar Pathfinder™

3. The conventional ladder size that will reach most single-story roofs is a(n) _____.
   a. 12-foot step ladder
   b. 12-foot extension ladder
   c. 18-foot step ladder
   d. 20-foot extension ladder

4. To gain access to the underside of an industrial roof 30 feet above the floor, the best choice is a 30-foot step ladder.
   a. True
   b. False

5. If a customer's house is using 1,500 kWh of power, and the calculations indicate that a 6.02 kW solar system is needed, roughly how many square feet will be required for the array?
   a. 150
   b. 602
   c. 1,500
   d. 6,020

6. The length of time a solar panel system can operate on battery power alone with no inputs is known as _____.
   a. autonomy
   b. solar threshold
   c. the battery bank
   d. smart metering

7. On motorized solar panel arrays, the panels are moved in order to _____.
   a. allow room for maintenance
   b. track the movement of the sun
   c. shed snow and ice
   d. adjust to prevailing wind conditions

8. Standoff mounting allows for better _____.
   a. cooling
   b. space for framing
   c. space for wiring
   d. water runoff

9. A measurement of 2½ inches equals ____.

   a. 25.6 millimeters
   b. 26.4 millimeters
   c. 63.5 millimeters
   d. 75.5 millimeters

10. Two 12 VDC panels wired in series will produce ____.

   a. 6 VDC
   b. 12 VDC
   c. 24 VDC
   d. 144 VDC

11. Sunlight intensity is measured in watts per square ____.

   a. inch
   b. foot
   c. meter
   d. millimeter

12. The site information needed to obtain a sun path chart from the University of Oregon Solar Radiation Monitoring Laboratory is the ____.

   a. area code and latitude
   b. zip code and time zone
   c. latitude, longitude, and time zone
   d. zip code, latitude, longitude, and time zone

13. The vertical lines on a Solar Pathfinder™ sun path chart represent ____.

   a. elevation
   b. altitude
   c. months
   d. hours

14. If solar panels are to be installed out in a field, any underground utilities must be located and ____.

   a. moved
   b. marked
   c. bypassed by 12 feet
   d. temporarily disconnected

15. The size of most charge controllers is ____.

   a. 4" by 4"
   b. 12" by 24"
   c. 24" by 24"
   d. 24" by 36"

Michael J. Powers

Corporate Safety Training Director
Tri-City Electrical Contractors, Inc.
Altamonte Springs, FL

Mike Powers has done everything in energy from wiring a kennel in the midst of a pack of barking dogs to working on solar panels. He has had a varied and rewarding career that has kept him interested in his work and enjoying life.

Who inspired you to enter the industry?
You could say the smell of hot grease inspired me. My first job was flipping hamburgers in a fast food restaurant, and it didn't take long till I was ready for a change. My dad was an electrician, and that looked like a much better career choice.

Tell us about your apprenticeship experience.
The program I apprenticed in was a good one. The electricians I worked under had a broad range of experience, so I was prepared for many different situations. I worked all over, in everything from kennels to colleges. I even worked in a fast-photo print shop once.

What positions have you held and how did they help you to get where you are now?
After I completed my apprenticeship, I became a licensed electrician, a job-site superintendent, a master electrician, and I'm currently a corporate safety and training director. The electrical theory taught in apprenticeship school made a big difference. I learned a lot studying for my licensing exams—those things gave me a good understanding of not just how, but why certain things were done. Then there was the practical experience of being on the job for over thirty years and seeing the kinds of things that were needed. These different kinds of experience gave me a good foundation for my current job in training. I'm now on the solar photovoltaic panel for NCCER's new program. It's stimulating to be in on something new and be part of a leading-edge technology.

What do you enjoy most about your work?
Problem-solving and people. Both are interesting and both can be a challenge.

Do you think training and education are important in construction?
If you want to have a career, it's not a choice. Our industry is changing and becoming more demanding every day, and you won't get anywhere standing still. The more you learn, the better you'll be prepared when an opportunity comes along.

Have you ever witnessed an accident that better training might have prevented?
I watched a carpenter step into a hole in a deck on the second floor. He stepped onto a 4 by 4 which he assumed was supported from below by a jack. He was not wearing any fall protection and the wood was toenailed in place, not braced. The board fell and he went with it. He suffered serious injuries. Because of the sand on the job site, several dozen workers had to push the ambulance to the building to take him to the hospital. Better training—and stricter enforcement of safety rules—might have made a difference.

How has training impacted your life and your career?
It's made me confident that I can do whatever comes my way. And it's given me the chance to move into new positions, make a better living, and expand my interests. It's kept me from getting stale.

How do you define craftsmanship?
Just like I define character—it's doing your job correctly, to the best of your ability, in a safe, productive manner, whether anyone else knows about it or not. It's personal.

SITE SURVEY CHECKLIST

SITE SURVEY CHECKLIST
☼ **GENERAL INFORMATION**
Date of Survey:
Site Name:
Contact Name:
Site Street Address:
City:　　　　**State:**　　　　**Zip:**　　　　**Country:**
Phone: (　　)　　　　　　**Fax:** (　　)
Email:

1. ROOF OR OTHER ARRAY MOUNTING SURFACE	
Check boxes or specify in the blank for items below.	
1.01	**Type of Roof Material or Mounting Surface (Specify)**
1.02	**Roof or Mounting Surface Condition**
1.03	**Age**
1.04	**Supporting Structure (e.g. roof trusses)**
	☐　Accessible
	☐　Adequate Strength
1.05	**Roof or Mounting Surface Slope (e.g., 5/12, flat)**
1.06	**Area (Sq. ft.)**
	- Azimuth Direction (degrees E or W of true South)
	- Eave Height (ft.)
	- Ridge Height (ft.)
1.07	**Accessibility to Proposed Array Location**
	☐　Easy
	☐　Moderate
	☐　Unacceptable
1.08	**Potential for Shading Proposed Array**
	☐　None
	☐　Slight
	☐　Unacceptable

2. INVERTER, UTILITY ACCESS, BATTERIES AND ENGINE-GENERATOR (AS APPLICABLE)	
2.01	**Proposed Inverter Location (Specify)**
2.02	**Accessibility to Proposed Inverter Location**
	☐　Easy
	☐　Moderate
	☐　Unacceptable
2.03	**Proposed Battery Location (Specify, if applicable)**
2.04	**Accessibility to Proposed Battery Location**
	☐　Adequate Ventilation
	☐　Adequate Location
	☐　Accessible
2.05	**Proposed Engine-Generator Location (Specify, if applicable)**
	☐　Adequate Ventilation
	☐　Adequate Location
	☐　Accessible

☼ **RECOMMENDATION**
Check the appropriate box below.

102A01A.EPS

Copyright © Florida Solar Energy Center

☐ Approve site for system installation

☐ Do not approve site for system installation (If site not approved, specify reasons for rejection below:)

☀ SURVEY REVIEWER INFORMATION

Name:

Organization:

Signature: **Date:**

Please list other committee members reviewing this design:

Name	Organization

SKETCH ROOF AREA AND PROPOSED ARRAY LOCATION (OR ATTACH ON A SEPARATE PAGE)

Available Roof Area (sq. ft.)

102A01B.EPS

Copyright © Florida Solar Energy Center

Trade Terms Introduced in This Module

Magnetic declination: The angle between magnetic north, as indicated by a compass needle, and true north (the North Pole).

Pitch: The ratio of the rise of a roof to its span expressed as a simple fraction.

Additional Resources

This module presents thorough resources for task training. The following resource material is suggested for further study.

AMtec Solar (combiners) website: www.amtecsolar.com.

Electric Power Glossary of Terms: www.osha.gov.

Florida Solar Energy Center website: www.fsec.ucf.edu.

National Geophysical Data Center website: www.ngdc.noaa.gov.

National Oceanic and Atmospheric Administration website: www.noaa.gov.

Solar Pathfinder™ website: www.solarpathfinder.com.

Solar Power Calculator website: www.findsolar.com.

Solar Source Institute website: www.solarsource.net.

Surrette/Rolls Battery website: www.surrette.com.

University of Oregon Solar Radiation Monitoring Laboratory website: solardat.uoregon.edu.

Figure Credits

Solar Pathfinder, Module opener, Figures 4 (photos) and 25–30

Topaz Publications, Inc., Figures 1 (long reel-style measure), 2, 7, 8, 14, 15, 17, and 19 (snap-type seal)

U.S. Tape Company, Figure 1 (measuring wheel)

The Stanley Works, Figure 3

Genie Industries, Figures 6 and 18

©2010 Photos.com, a division of Getty Images. All rights reserved, Figures 11 and 19 (shingle roofing)

Dovetail Solar and Wind, Figure 12

ProtekPark Solar, Figure 13

ATAS International, Figure 19 (machine-crimped panels)

Maruhachi Ceramics of America, Figure 19 (ceramic Spanish tiles)

DaVinci Roofscapes LLC. For more information about DaVinci's family of products, call 1-800-328-4624 or visit www.davinciroofscapes.com., Figure 19 (simulated wood shingle roof)

Unirac, Inc., Figure 21

Sunpath charts are generated from a website program of the University of Oregon Solar Radiation Monitoring Laboratory, www.solardat.uoregon.edu/SunChartProgram.php, Figures 23 and 24

Sun Banks Solar, Figure 31

AMtec Solar, Figure 32

Outback Power Systems, Figure 33

Surrette Battery Co. Ltd., Figure 34 (FLA battery)

Concorde Battery Corporation, Figure 34 (AGM battery)

Midnite Solar, Figure 35

Fronius International GmbH, Figure 36

SMA Solar Technology AG, Figure 37

Itron Inc., Figure 38

Xcel Energy, Figure 39

Florida Solar Energy Research Center® (FSEC®), a research institute of the University of Central Florida, Appendix

NCCER CURRICULA — USER UPDATE

NCCER makes every effort to keep its textbooks up-to-date and free of technical errors. We appreciate your help in this process. If you find an error, a typographical mistake, or an inaccuracy in NCCER's curricula, please fill out this form (or a photocopy), or complete the online form at **www.nccer.org/olf**. Be sure to include the exact module ID number, page number, a detailed description, and your recommended correction. Your input will be brought to the attention of the Authoring Team. Thank you for your assistance.

Instructors – If you have an idea for improving this textbook, or have found that additional materials were necessary to teach this module effectively, please let us know so that we may present your suggestions to the Authoring Team.

NCCER Product Development and Revision

13614 Progress Blvd., Alachua, FL 32615

Email: curriculum@nccer.org
Online: www.nccer.org/olf

❏ Trainee Guide   ❏ Lesson Plans   ❏ Exam   ❏ PowerPoints   Other _____

Craft / Level: _____   Copyright Date: _____

Module ID Number / Title: _____

Section Number(s): _____

Description: _____

_____

_____

_____

_____

Recommended Correction: _____

_____

_____

_____

_____

Your Name: _____

Address: _____

_____

Email: _____   Phone: _____

System Design

57103-11

Trainees with successful module completions may be eligible for credentialing through NCCER's National Registry. To learn more, go to www.nccer.org or contact us at **1.888.622.3720.** Our website has information on the latest product releases and training, as well as online versions of our *Cornerstone* magazine and Pearson's product catalog.

  Your feedback is welcome. You may email your comments to **curriculum@nccer.org,** send general comments and inquiries to **info@nccer.org,** or use the User Update form at the back of this module.

 V.2 11/13

Objectives

When you have completed this module, you will be able to do the following:

1. Identify appropriate system designs and array configurations based on user loads, customer expectations, and site conditions.
2. Determine the size and capacities for major system components based on user load, desired energy production, autonomy requirements, and costs.
3. Determine the PV panel layout, orientation, and mounting method for optimum system production and integrity.
4. Determine the ampacity requirement for all components and wiring of the PV system.
5. Select the appropriate conductor types and sizes for each portion of the electrical circuit.
6. Identify the appropriate size, rating, and location of required overcurrent protection and power disconnect devices.
7. Determine the appropriate size, rating, and location for bonding, grounding, and surge suppression.

Performance Task

Under the supervision of the instructor, you should be able to do the following:

1. Given a completed site assessment, design a grid-connected PV system.

Prerequisites

Before you begin this module, it is recommended that you successfully complete *Core Curriculum* and *Solar Photovoltaic Systems Installer*, Modules 57101-11 and 57102-11. It is also suggested that you shall have successfully completed the following modules from the Electrical curriculum: *Electrical Level One*, Modules 26101 through 26111; *Electrical Level Two*, Modules 26201, 26205, 26206, and 26208 through 26211; *Electrical Level Three*, Modules 26301 and 26302; and *Electrical Level Four*, Modules 26403 and 26413.

Trade Terms

Absorption stage	Equalization stage	Shunt controller
Battery bank	Equipment grounding	Single-stage controller
Bulk charge stage	Float stage	Solar combiner
Bulk voltage setting	Ground fault protection	Sulfation
Days of autonomy	Interconnection agreement	System grounding
Diversion controller	Low-voltage disconnect (LVD)	

Note: *NFPA 70®*, *National Electrical Code®*, and *NEC®* are registered trademarks of the National Fire Protection Association, Inc., Quincy, MA 02269. All *National Electrical Code®* and *NEC®* references in this module refer to the 2011 edition of the *National Electrical Code®*.

Contents

Topics to be presented in this module include:

Figures and Tables

1.0.0 INTRODUCTION

Before the solar PV system design process can begin, the designer must first decide which configuration is appropriate, based primarily on the site assessment and budget considerations. There are a number of factors that determine the final choice. One major factor will be the desires of the owner. In some cases, an experienced designer may recommend a particular approach based on site assessment data, the electrical load, and other relevant conditions. But in the end, owners may choose a different approach due to budget restraints or, in other cases, due to their desire to invest more deeply for personal or environmental reasons. Their choice may be at odds with the typical choices others make, but it is the duty of the designer to fully discuss all aspects of the possible configurations to help the owner make an informed choice. A sensational design and installation that does not meet the owner's expectations must be considered a poor result.

The basic system configurations were reviewed in *Introduction to Solar Photovoltaics*. They include:

- *Stand-alone systems (sometimes referred to as off-grid systems)* – This category includes day-use systems, where no power storage is used, and systems that employ batteries to store power for use when the PV array is not producing power. Stand-alone systems can also be mated with other forms of off-grid energy, such as generators or wind turbines. Both AC and DC loads can be accommodated.
- *Grid-tied (often referred to as grid-connected or utility-interactive) systems* – This category includes any system which is tied to the local utility power system. Grid-tied systems can also include batteries for power storage, generators as a third source of power, etc. Some grid-tie systems incorporate the means to sell excess power produced back to the utility.

Some design tasks for the various systems are similar in nature. In all cases, there are solar panel and array configurations to be chosen, and wiring to be sized. For stand-alone systems, no grid interface is involved and those design tasks are eliminated.

2.0.0 STAND-ALONE SYSTEM DESIGN

Stand-alone systems must be designed to provide 100 percent of the desired or required energy needs, while grid-connected systems can be sized to provide virtually any amount. Any amount of power generation, up to and beyond 100 percent of the projected energy consumption, can be incorporated into the grid-connected system design to satisfy the owner's goals. Power production beyond what the site requires can be sold back to the utility if this is an objective of the owner. The focus of this module is on the stand-alone system application.

Although simple stand-alone systems can be installed that do not incorporate any sort of battery for power storage, they are obviously without power in the absence of sunlight. Since stand-alone solar PV systems without storage capability have little practical application in dwellings and many other situations, a system with battery storage is the focus of the example system. The component layout of a simple stand-alone system can be seen in *Figure 1*.

The basic steps to design a stand-alone system are as follows:

Step 1 *Determining the electrical load* – This step is generally completed as part of the site assessment process. Once the existing electrical loads are determined, some detailed discussion with the owner regarding strategies to minimize the load is in order. Every watt of needed power adds to the system investment. There are many possibilities to consider, such as more energy-efficient lighting, refrigeration, and heating/cooling systems. These three common loads represent a huge portion of the electrical needs in a typical home, and therefore also represent the best opportunities for energy savings. Before determining the final electrical load then, the designer or site assessor must examine all potential additions or subtractions that may not presently exist.

Step 2 *Selection and sizing of batteries* – A variety of chemicals and elements can be combined to make batteries. Some combinations are low cost but provide low amounts of power, while others can store significant amounts of power at a proportionately high cost.

Step 3 *Selection and sizing of the PV panels* – There are numerous choices in solar panels today, and even more choices enter the market on a regular basis.

Step 4 *Selection of the controller* – The controller provides an interface between the solar array and the batteries, providing protection against overcharging through voltage regulation. A number of other important features are also available, such as protecting the batteries from overdischarge.

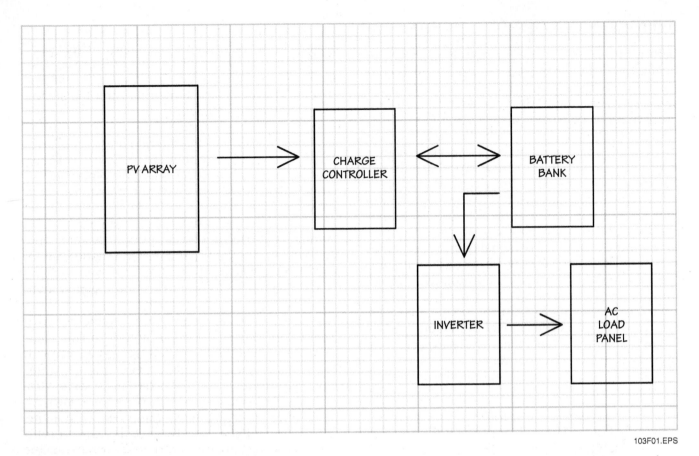

Figure 1 Stand-alone system component diagram.

Step 5 *Selection and sizing of the inverter* – The inverter provides the interface between the DC power generated by the array as well as the energy stored in batteries, and the AC electrical system which powers all conventional electrical loads. Some inverters can handle battery charging duties as well, eliminating the need for a separate charge controller.

Step 6 *Selection and sizing of all interconnecting wiring and protective devices* – The *National Electric Code*® (NEC®) incorporates standards for PV systems in special sections devoted specifically to such systems. In addition, common standards for wire sizing and protective devices also apply.

These basic design steps will first be examined in a simple, stand-alone PV system design using a small, remote dwelling as an example. This example will also incorporate more specific decisions to be made in the design process which go beyond the simple steps above. Forms have been created for use in the component selection process and are found in *Appendix A* in the back of this module.

2.1.0 The Electrical Load

For the example home discussed in this module, the owner desires a stand-alone system. Consideration has already been given to the use of energy-efficient appliances and lighting during the site assessment. There is no cooling system which requires electrical power, and heating is accomplished by alternative forms of energy.

The following information gathered during the site assessment is offered in *Table 1* as a basis for the design electrical load.

The design AC load wattage is the sum of the applied loads (2,940W). The total AC average load is 4,239Wh per day. It is important to note that this example is a relatively simple one, and the loads listed here are quite general in nature. It cannot be stressed enough that the site assessment must be extremely thorough, and loads which do not presently exist may be introduced later. Also in today's homes, many tiny but consistent and continuous electrical loads exist, such as the small amount of power consumed by a DVD player/recorder when the system is not in use. These loads are often referred to as phantom loads and should be addressed in the assessment, regardless of how small.

NCCER — *Contren® Learning Series* 57103-11

Table 1 Electrical Load Data for a Stand-Alone PV System

Load	Qty	Volts	Amps	AC Watts	Hrs/Day	Days/Week	7 Days/Week	AC Wh*
Microwave	1	120	9.5	1,140	.2	7	7	228
Washer	1	120	9.9	1,188	2	1	7	339
Refrigerator	1	120	2.6	312	6	7	7	1,872
Lighting (CF)	12	120	2.5	300	6	7	7	1,800
							Total AC Average Daily Load =	4,239

*Qty × volts × amps = AC watts × hrs/day × days/week ÷ 7 days/week = AC Wh

A number of components suffer losses of electrical efficiency in their operation. Inverters, for example, often operate at a 90 percent level of efficiency—roughly 10 percent of the power input is lost to heat and other factors. These losses must be considered mathematically as needed throughout the process. However, these calculations of inefficiency should not be considered fudge factors or cushions—they represent realistic expectations of actual component performance. If the owner desires an extra measure of design capacity for future use, or simply wishes to ensure that the system output not only meets but exceeds the AC average daily load, additional capacity may be required. Calculations to recognize inefficiency do not contribute to this end. This example uses a daily load of 4.239kWh.

2.2.0 Battery Selection

In order to select and determine the number of batteries required for this system, several other design decisions must be made first. Begin with the selection of the DC system voltage.

Since most flooded lead acid (FLA) batteries suitable for solar PV use are rated at 12 VDC, one could assume that this would be the voltage of choice for the DC side of the system. For very small system loads and applications, it would be a reasonable choice. However, voltage and current are directly related and the current for an operating electrical load can be cut in half by doubling the voltage. Current is the primary factor in wire sizing, and wire size has a great deal to do with its cost. In systems providing power to significant loads, operating at low voltages on the DC side can significantly increase the wire size. Resistance must be minimized in the wiring through proper sizing, thus practical designs may demand that system DC voltage be increased above 12V to reduce current flow and, in turn, wire sizes.

Since batteries can be combined in series to increase their voltage, it follows that battery banks can easily be created incrementally to achieve the desired bank voltage. Systems are generally installed based on configurations of 12V, 24V, or 48V. These values also fall in line with available inverters, to be discussed later in this module.

A quick calculation to determine the approximate current flow in the example can be done using power divided by voltage equals amps. If using a 12V system, then 4,239W ÷ 12V = 353A. 12V is not a good choice since a massive wire size would be required to carry this current. Using 48V in the same equation results in a current flow of 88A. Quadrupling the system voltage reduces the current flow to 25 percent of the original value. This 88A value is a much more reasonable value to work with for wire sizing and overcurrent protective devices. Increasing the DC voltage of the system further is, of course, electrically possible. However, higher-voltage systems present greater safety hazards, and components such as inverters which operate at DC voltages beyond 48V are expensive and more difficult to obtain. The *NEC*® requires that systems operating at greater than 48 VDC be equipped with an electric means of disconnect to separate the series strings of batteries into groups of 48V or less, allowing for safer maintenance activities. Therefore, a 48V system will be chosen for the example system.

2.2.1 Days of Autonomy

The next issue is related to the desired number of days of autonomy. In solar PV systems, autonomy is related to the time period when all critical system loads continue to operate without the batteries being recharged. Obviously some level of autonomy in a stand-alone system is required if power is needed consistently, since the PV system will not generate power in darkness and power generation can be severely limited for significant periods of time by weather conditions. In some areas, the weather may be relatively clear for several weeks, or even months, at a time. As the

System Design

seasons change though, this same region may be shrouded in fog for significant periods of time. All climate-, weather-, and sun-related information should be taken into consideration, along with the owner's expectations, when determining the days of autonomy for battery sizing.

Begin with a calculated approach, based on the premise that a longer period of autonomy is indicated in areas where solar availability is limited. Examine the peak hours of sun per day for the given location, expressed mathematically as $(kW/m^2)/$ day, and for the month with the lowest solar insolation of the year. For a location such as Fairbanks, AK, this number would be $0.3 (kW/m^2)/$day. However, a location like Las Vegas, NV enjoys a value of 4.9. Such differences have a significant impact on the days of autonomy chosen for the design. *Table 2* provides one set of recommendations.

If a backup source of power is involved, such as a generator, then only two to three days of autonomy are normally considered, and possibly even less.

There are other factors to consider when contemplating the days of autonomy for a given system. Consider this scenario, with a given system which has been designed for three days of autonomy. It would generally be sized to provide a full day of the energy needed. The system is started, batteries are charged fully, and then the electrical load is imposed. As long as excellent solar insolation conditions remain, the batteries will charge and discharge in the upper 30 to 35 percent of their capacity range. But foul weather sets in, rendering the array virtually useless. As the three days pass, the batteries eventually reach their low-voltage disconnect (LVD) value. This is the value used for control settings to disconnect the batteries from the load, protecting them from damage due to extreme discharge. This setpoint will generally be at the 20 to 25 percent level of the batteries' capacity. As the sun returns and array production begins, the batteries will charge sufficiently to power the load each day, but never recover fully since the daily load and the daily charge are roughly equal. The result is that, once this condition is first encountered, the system is unable to properly recover unless the load is reduced or eliminated for some period of time. In theory, eliminating the load for three days will fully recharge the batteries. Likewise, reducing the load by 50 percent for six days could also accomplish the same goal. This high degree of hands-on system management may not be available, especially in remote applications. The life of batteries left operating under the above conditions (in the low end of the charging capacity range) will be significantly shortened.

This operating phenomenon should encourage designers and system owners to minimize the days of autonomy designed into a battery system as much as practical, resulting in smaller battery banks. Although a smaller bank obviously reaches the LVD point quicker than a larger bank, there is virtually no difference in functionality (the phenomenon will eventually occur anyway), and the smaller battery bank has a better chance of recovering fully at a faster pace. PV system batteries are expensive, and protecting their ability and longevity should begin in the design phase.

The example home is in Las Vegas, NV, which has a peak sun hour per day value in the lowest month of 4.9. Per the chart above, the system would normally provide five days of autonomy. However, the 4.9 $(kW/m^2)/$day value is above the 4.5 value in the chart, and the owner is comfortable with the concept of cutting back household energy needs on the rare occasions of extremely poor weather. Therefore, this example system will be designed for only four days of autonomy to help reduce the system battery bank size and cost.

2.2.2 Battery Operating Temperatures

Battery temperatures have a significant impact on their performance. The capacity of a battery decreases at lower temperatures, and increases at higher temperatures. However, battery life increases at lower temperatures and decreases at higher temperatures. Therefore, it is important to select an environment for the battery bank that is as reasonable in temperature as possible throughout the year to avoid performance penalties of one type or another. Further, it is important to consider the location before making a final battery selection. Battery manufacturers rate the performance of their batteries at 77°F (25°C). Deviations from this value in the installation (and some deviation is quite likely) result in altered performance from the battery specifications.

The designer needs to consider temperature in the selection process if it is known that the storage area will be excessively hot or cold, derating the

Table 2 Recommended Days of Autonomy Based on Solar Insolation

kW/m²/day	Recommended Days of Autonomy
4.5+	5
3.5 to 4.5	6
2.7 to 3.5	7
2.0 to 2.7	8
< 2.0	Up to 14

capacity or voltage settings accordingly. For the purposes of this exercise, assume that the average temperature of the battery bank environment will be 77°F year-round. In reality, the temperature will likely vary roughly ±20°F, affecting the battery bank capacity and life accordingly.

Note that the battery bank should be housed in a sturdy, noncorrosive housing (*Figure 2*) when possible. Per *NEC Section 690.71(D)*, systems operating above 48V must be stored in a non-conductive case, but it is always a good idea at any voltage. Provisions for battery ventilation must also be made, which is addressed during the installation phase. In some cases, depending upon the number of batteries involved, several housings may be required. The housing can be insulated or uninsulated, further providing the designer an opportunity to control battery temperature, especially in cold climates. Under no circumstances should other electrical devices or components be installed inside, or even near, the battery bank.

2.2.3 Battery Capacity

Nominal battery capacity is rated in amp-hours (Ah). Generally speaking, a battery rated at 200 Ah can deliver 2A for 100 hours, or 5A for 40 hours. Just as batteries can be connected in series to increase system voltage, they can be wired in parallel to increase the total Ah capacity.

In reality though, actual battery capacity is affected by several factors. A rapid rate of discharge, for example, results in reduced capacity, while a slow discharge cycle results in increased capacity. Battery ratings are then provided based on a given discharge rate. A battery discharge rate of C20 means that its capacity is based on a 20 hour discharge cycle—quite typical for solar PV system batteries. A rating of C5 would mean

103F02.EPS

Figure 2   Field-constructed PV system battery enclosure.

the capacity is based on a five hour discharge cycle. The designer's knowledge of the system application and usage help determine what rating is appropriate, but batteries for residential solar applications are generally selected from their values at 20 to 24 hour discharge cycles.

The depth of discharge must also be considered in the battery selection. Depth of discharge is a measure of how deep a battery's capacity will be drawn down. The maximum level of drawdown is generally considered to be 80 percent of capacity, but this is not a value that is practical or beneficial to consider as a design target. FLA batteries should never be fully discharged, as they quickly suffer a decrease in their output voltage. The overall life of the battery is severely decreased if it is allowed to cycle this deeply. A battery drawn down to 50 percent of its capacity repeatedly lasts roughly twice as long as a battery consistently drawn down by 80 percent.

Batteries which are discharged only 10 or 20 percent have even longer lives. However, designing a battery bank to achieve this level of performance substantially increases the number of batteries that are required, create charging issues, and greatly increase the cost of the installation. If the bank is too large in capacity, when compared to the solar array charging capacity, the batteries may not be charged to full capacity in the available array production period. This shortens battery life as the result of sulfation. In most applications, a 50 percent depth of discharge is considered a good design point as it balances battery life, initial cost, and available capacity.

2.2.4 Power Inverter Efficiency

One additional piece of information needed to begin the battery selection calculation is the inverter's projected operating efficiency. Inverter selection is covered in more detail later in this module, but its efficiency must be factored into the battery selection since it represents a loss in usable battery power. As a general rule, inverters operate around 90 percent efficiency. This figure can be used in initial calculations, but the designer should return to the battery selection process after selecting the inverter and determining its actual efficiency to ensure accurate calculations.

2.2.5 Battery Selection Calculations

Before beginning the calculation, review the information gathered thus far:

- The average AC daily current load will be 4,239 watt/hours.
- There are no DC loads to be considered.

- The inverter efficiency is expected to be 90 percent, or 0.90.
- The DC system voltage will be 48 VDC.
- The desired number of days of autonomy is 4 days.
- The design discharge limit of the batteries is 50 percent, or a 0.50 multiplier.
- The batteries will be stored in an area where their temperature will remain near 77°F.

The battery selection process incorporates three simple equations. The first equation is used to determine the average amp-hours (Ah) per day needed. It is calculated as follows:

[(AC watt/hour average daily load ÷ inverter efficiency) + DC average daily load] ÷ DC system voltage = average Ah/day

Solving the equation with the example values results in:

$$[(4{,}239 \div 0.90) + 0] \div 48\ VDC = 4{,}710 \div 48 = 98\ average\ Ah/day$$

Note that the final result has been rounded up to the next highest whole number.

Next, determine the number of batteries that will be wired in parallel to provide the required capacity:

Number of batteries in parallel = [(average Ah/day × days of autonomy) ÷ depth of discharge limit] ÷ selected battery Ah capacity

Note that one piece of information needed to solve the above equation has not yet been determined—the selected battery capacity. Generally, solar installers have a particular line of batteries that they work with and recommend. In some cases, experience will tell the designer what capacity of individual battery is likely to be the best choice. For this example, first examine the result of the first part of the equation:

Battery capacity = (Average Ah/day × days of autonomy) ÷ depth of discharge limit

$$Battery\ capacity = (98 \times 4\ days) \div 0.50 = 392Ah \div 0.50 = 784Ah$$

This is the capacity required when wiring the batteries in parallel. Now refer to a list of batteries from one manufacturer (*Table 3*).

Using the 212Ah model, it would take 4 batteries in parallel (4 × 212Ah = 848Ah) to reach the 784 needed. This exceeds the system requirements. Using the 258Ah model, three batteries are required to obtain 774Ah, slightly short of the goal. In this example, the 212Ah model will be a reasonable choice. This value is plugged into the previous equation:

Table 3 Typical Battery Specifications

Product Model	Nominal Capacity Ah @ 77°F, Down to 1.75V per Cell
B-1040	104
B-1080	108
B-2120	212
B-2580	258

$$[(98 \times 4\ days) \div 0.50] \div 212Ah = (392 \div 0.50) \div 212 = 4\ batteries\ in\ parallel$$

Note that the actual answer of 3.7 batteries has been rounded up to 4 batteries.

Finally, determine the total number of batteries needed in series by dividing the DC system design voltage by the battery voltage to find the number of batteries required in series. Use the following equation:

Batteries required in series = DC system voltage ÷ battery voltage

Batteries required in series = 48V ÷ 12V batteries = 4 batteries in series

The total number of batteries required for the battery bank can now be determined by multiplying the number in series by the number in parallel:

Total batteries required = batteries required in parallel × batteries required in series

Total batteries required = 4 batteries in parallel × 4 batteries in series = 16 batteries

This system will require a total of (16) 12V, 212Ah batteries to provide the needed capacity for four days of autonomy, at the design electrical load. See *Figure 3*.

2.3.0 Solar PV Panel Selection

Although the calculations involved in the selection of solar panels are not difficult, the process does require a thorough understanding of panel specifications to understand which panel to select and to be able to predict their performance when mated with the other system components. Although some of that information is explained here, it is also important to note that solar panels do not perform equally, nor do they always perform to their manufacturer's specifications. This complicates the selection process significantly. Solar panel performance can also degrade over time. Complicating the process even further are the many different factors that affect their field performance, including panel temperature, sunlight intensity, and load resistance. When a

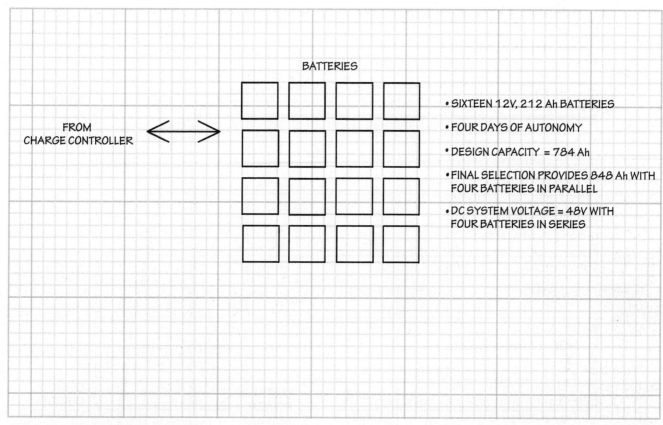

Figure 3 Battery selection results.

BATTERIES

- SIXTEEN 12V, 212 Ah BATTERIES
- FOUR DAYS OF AUTONOMY
- DESIGN CAPACITY = 784 Ah
- FINAL SELECTION PROVIDES 848 Ah WITH FOUR BATTERIES IN PARALLEL
- DC SYSTEM VOLTAGE = 48V WITH FOUR BATTERIES IN SERIES

FROM CHARGE CONTROLLER

103F03.EPS

designer and/or installer suspects that an array is not performing up to its specifications, proving the theory with so many other potential factors involved can be challenging.

Most designers soon become affiliated with a particular panel manufacturer after entering the field. The best starting point is to draw from the experience of others. One important feature to the owner should be the warranty offered. Panels which offer shorter warranties are often expected by the owner to fail sooner, regardless of whether or not this assumption is true. Perception in the quality of the product is as important as reality here, as many years may go by before the truth is revealed. Owners that make substantial investments in their PV system will feel more comfortable with warranties in the 25-year range.

2.3.1 PV Panel Performance Basics

Understanding the electrical performance of the individual solar panels begins with a review of the performance curve, where current output is compared to voltage output. This curve is known as the I-V curve (*Figure 4*).

These I-V curves are obtained by exposing the cell to a constant level of light, while maintaining a constant cell temperature, varying the resistance of the load, and measuring the current

that is produced. It is very important to note that standard test conditions are 77°F (25°C) cell temperature, 1kW/m² irradiance (sometimes referred to as 1 peak sun hour), with 1.5 ATM spectral conditions. Using standard test conditions does help the designer to compare panel performance, but in reality these conditions may rarely exist at the installation site. Since the panels are located in bright sun, temperatures often drift higher. The panel will also experience the 1kW/m² irradiance only periodically. Since the designer must account for actual conditions, it is imperative that the I-V curve is understood and utilized.

The curve identifies the possible combinations of current and voltage output of a photovoltaic panel. A PV panel produces maximum current when there is no resistance in the circuit, essentially when there is a short circuit between its positive and negative terminals. This maximum current is known as the short circuit current (I_{sc}). When the panel is shorted, the voltage in the circuit is zero. Conversely, the maximum voltage occurs when the circuit is broken. This is called the open circuit voltage (V_{oc}). Under this condition the resistance is infinitely high, and there is no current, since the circuit is incomplete.

Note the points identified in the graph as short circuit current and open circuit voltage. These two

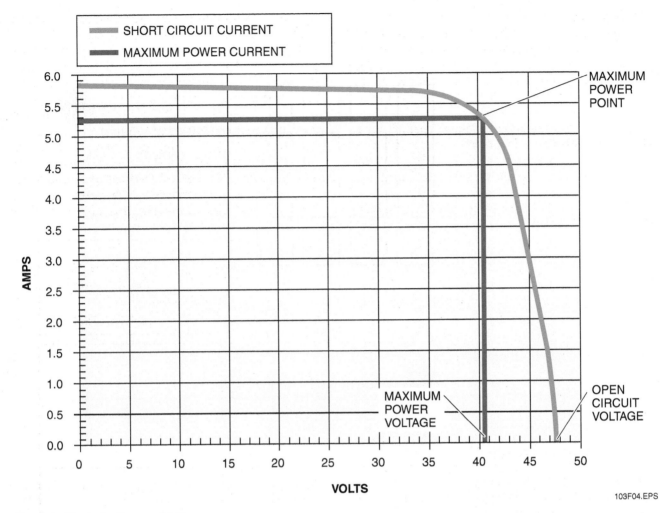

Figure 4 Typical solar panel I-V curve.

extremes in load resistance (open circuit versus short circuit) and the whole range of conditions in between are depicted on the I-V curve. Current, in amps, is plotted on the vertical axis and voltage, in volts, is plotted on the horizontal axis. The power available from a photovoltaic device at any point along the curve is simply the product of current and voltage at that point and is expressed in watts. At the short circuit current point, the power output is zero, since the voltage is zero. At the open circuit voltage point, the power output is also zero, but this time it is because the current is zero.

The panel can function over a wide range of voltages and currents. The panel delivers its best performance and highest efficiency at the knee of the curve. This is the point where the panel delivers maximum power. Remember that power is the product of voltage times current. Therefore, on the example I-V curve, the maximum power point occurs where the product of current times voltage is at its maximum value. This point is the intersection of maximum voltage and maximum current. Again, no power is produced at the short-circuit current with no voltage, or at open-circuit voltage with

no current. On the graph, the maximum power point is at 40V and 5.3A.

Panel specification labels (*Figure 5*) contain information that is taken from their I-V curve. Standard test conditions may also be listed.

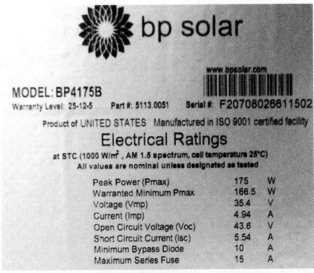

Figure 5 Solar panel specification label.

In *Figure 6*, the graph has been modified to show the impact of changes to the standard test conditions. As the irradiance is reduced below 1 peak sun hour, the resulting performance line slides down the vertical axis. As higher temperatures are experienced, the line slides toward zero on the horizontal axis. Higher panel temperatures result in reduced voltage output but an insignificant change in current output. Reduced sunlight results in a reduction of current output, but does not affect voltage significantly. Note that the shape of the graph changes very little—as temperature or sunlight quality change, the basic

shape is maintained. The graph shape is simply repositioned in relation to the two axes.

In addition to considering the impact of temperature and irradiance, the designer must also consider any potential for shading to occur at the site. A thorough examination of shading potential is an integral part of the site assessment, and the information gathered should be reviewed carefully during the design phase. Due to the details of panel construction, the performance of a solar panel can be reduced by an astounding 75 percent if only one cell of a panel becomes completely shaded. Total shading of 3 individual

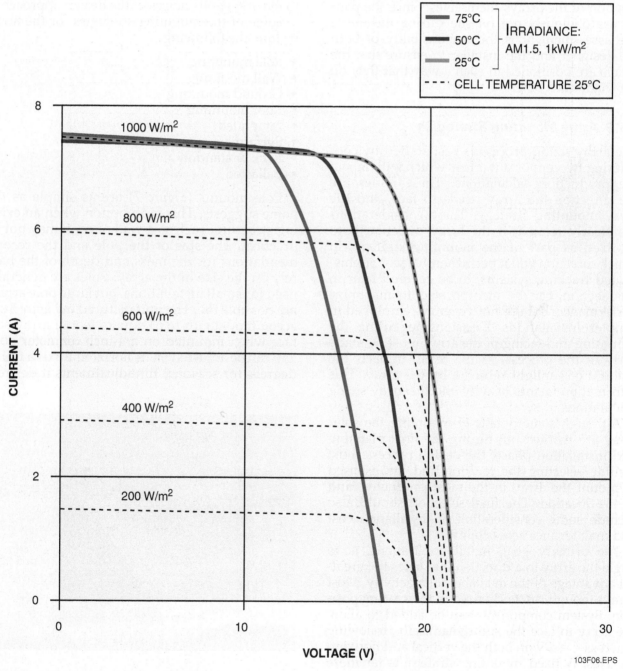

Figure 6 I-V curve at varying conditions.

cells can result in a 90 to 95 percent loss of performance. Panels that incorporate internal bypass diodes (the electrical version of a check valve) can reduce these loss figures somewhat, but certainly not enough to discount shading.

For this reason, the designer should avoid placement of a solar array in a location which could experience any level of shading between 9 AM and 3 PM, when maximum performance is needed and anticipated. If it cannot be avoided, then more panels in the array are necessary to achieve the desired results. Information regarding shading comes forward from a thorough site assessment, with the exception of one factor—the position of the arrays themselves. Since the panels are obviously not on site during the initial site assessment, it is the responsibility of both the designer and the installer to ensure that the arrays are positioned in such a way that they do not shade each other.

2.3.2 Array Mounting Strategies

The array sizing process is best started by considering two important issues which will impact the production calculations. These issues are the direction the array needs to face, and the array mounting strategy. These considerations are clearly related. If the array direction is to be fixed as part of the mounting strategy, in which direction will it permanently face? Sophisticated tracking systems, to be covered later in this section, can be incorporated to maximize performance, but are not generally employed in simpler installations. Decisions regarding the mounting and facing of the array must be made early in the process, as the result impacts the amount of sunlight received by the array. This information factors heavily into the array sizing calculations.

Although some details of mounting the solar array are worked out during the final planning and installation phase, the design process must include selecting the location and means used to mount the array in the most productive and secure location. The final selection should also include some consideration of installation cost and maintenance accessibility.

The primary goal, in all applications, is to angle the array in a direction that takes the greatest advantage of the available solar activity. Most arrays though, are fairly fixed in their facing position. System components can be added to allow the array to face the sun dynamically, following it across the sky on both the vertical and horizontal axes. A fixed mounting strategy is far more popular though, and generally more cost effective

at this time. If there is a desire to make up the difference in power production that a dynamic array would provide, a few extra solar panels can often be added to the array at a lesser cost to achieve the same result.

The site assessment provides the designer with information specific to the site. The facing of the structure, orientation of the roof, and any objects which could potentially shade the array must be considered carefully. If, for example, a home's angled roof faces east and west, then it may not be a practical location for the array. A flat roof though, offers an opportunity to orient the array at any angle. A detailed site assessment is critical to making good choices in the design approach.

Some of the mounting strategies for the array include the following:

- Pole mounting
- Wall mounting
- Ground mounting
- Roof mounting
- Integrated
- Surface
- Rack or standoff
- Ballasted

Pole mounts (*Figure 7*) are as simple as the name suggests. This is an option when an existing structure, such as a wall or roof, may not be practical. The size of the pole and the recommendations for the mass and depth of the base rely on the size of the array. Poles are generally used for small installations, but large pole arrays are possible too. The unit pictured in *Figure 7* can accommodate up to 65 square feet of array surface when mounted on a 4-inch diameter pole, and can be adjusted on its horizontal axis a full 90 degrees for seasonal tilt adjustments if desired.

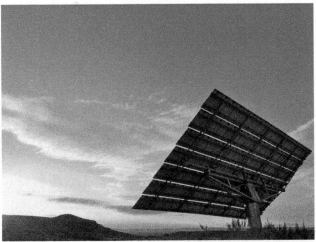

103F07.EPS

Figure 7 Pole-mounted solar array.

Several pole-mounted arrays can be installed in a cluster to create large systems. Accessibility for cleaning and maintenance should be considered carefully, but it is often easier than with other options. Pole mounts offer some significant advantages in many cases, since the array assembly can usually rotate on a vertical axis, while also having some adjustability on its horizontal axis. When multiple pole-mounted arrays are used, it is essential that they be positioned to avoid any chance of shading each other. Keep in mind that anchoring an array post in rock requires a very different approach than anchoring the post in sandy soil.

Wall mounts are limited in practical size, but can be used for small arrays. Assuming the wall mount application is fairly low, it provides good access for service and cleaning.

A number of different panel characteristics have been developed to allow for more aesthetically pleasing installations. Solar panels with some transparency can be used, as well as panels that are a particular color or a special shape. Arrays can even be used as a sort of Venetian blind for buildings, reducing solar gain on the facility while producing power.

Ground mounting systems are quite popular for many reasons and are limited in size only by the available ground space. Many solar panel manufacturers also offer factory-built racks for ground mounting. Better rack systems incorporate manual adjustments that allow for seasonal tilt changes. The terrain for ground mounts must be of reasonable integrity, and substantial anchoring systems are sometimes necessary to withstand wind loads. The integrity of the entire supporting structure is important to ensure the security of the very expensive array. Ground-mounted arrays provide for easy installation, maintenance, and cleaning. The only negative factor is often the loss of ground area where space is limited or of high value.

Roof-mounted systems vary widely. Integrated arrays are incorporated directly into the roof structure, usually along with the roof itself. But integrated arrays are not limited to the roof alone. For example, they can also serve as awnings over entrances. Integrated systems are best planned for when the building is being constructed, rather than retrofitted.

Surface mounting on a pitched roof (*Figure 8*) is popular and relatively simple. Today's commercially available mounting systems offer great flexibility and security, even when mounting to unique roof structures such as the one shown in *Figure 8*. The panels are generally supported just above the roof surface by a rack system, which allows for airflow under the panels to aid in their

On Site

BIPV Systems

Building integrated photovoltaics (BIPVs) generally refers to photovoltaic systems integrated with a buildings' original architecture. It is possible to install such a system in a completed building as well. Facade roof systems, for example, can be added after the fact. It is imperative that a great deal of care be given to the design process, and such systems can challenge even the best architect. Accessibility for service and cleaning, for example, can be quite difficult. Since the solar panels become an integral part of the structure itself, unusual stresses and loadbearing capability must be considered. In the application shown here, solar panels have been incorporated into the roof membrane being applied.

103SA01.EPS

Figure 8 Solar array mounted on a pitched roof.

cooling. The mounting rack is attached directly to the roof supporting structure, not to the roof sheathing. The pitch of the roof, determined during the site assessment, will determine if the mounting rack must be built to tilt the array to a different angle from that of the roof itself. Cleaning and service access must be considered carefully for pitched roof-mounted applications.

A flat roof offers an opportunity for a rack much like one that would be used for a ground installation (*Figure 9*). The structure can be anchored to the roof support members. Smaller arrays may also use ballast, such as sand bags, to hold them in position since penetrating an existing flat roof for anchorage often creates problems. High winds can be an obvious problem for ballast methods, but it is an acceptable practice in some applications. Local codes may prevail.

Solar tracking systems can add significantly to power generation, as much as 25 to 40 percent.

Tracking typically provides larger gains during the summer months. However, tracking systems add significantly to the installation cost, and also add to the list of items which can potentially fail or require periodic maintenance. In many cases, even more structural support is required since the repositioning of the array also changes wind load effects and structural balance. Tracking set-ups are either single-axis, tracking only the azimuth (east to west position of south), or double-axis to track both the azimuth and altitude.

Once one or two locations that offer the best potential for power generation are selected, consider the likely installation cost, maintenance, and service accessibility between the options. Remember that any shading, especially during peak production hours, can result in a major loss of production. As a final consideration, determine where the BOS components will be located. These include items such as the batteries, controllers, and inverters. One potential array location may be significantly closer to the BOS components than another, reducing the cost and complexity of interconnecting wiring.

2.3.3 Facing the Array

The direction a fixed array is faced is often at latitude, where a reasonable value of irradiance is found year-round. Since summer performance is better with the array directed at latitude –15°, and winter performance is better at latitude +15°, facing the array toward latitude offers the best year-round alternative. However, the designer must consider the details of the site assessment and the owner's desires. In some cases, the owner will be more concerned about array performance during a specific time of the year. For those projects, the array should be faced as indicated for best performance, and it is understood that performance during other times of the year will suffer as a result.

Figure 10 provides solar insolation data for the Las Vegas area, where the example home is located. An examination of the figures will confirm the performance gains and losses resulting from different array orientations at different times of the year. Since the owner does not have a specific time of year when the solar performance is more important than another, plan on installing the array facing toward latitude. At this juncture, a pole-mounted array would be the first choice to take advantage of its facing flexibility and ease of installation, but final details cannot be considered until the overall size of the array is known. With this style of mount, the owner can choose to adjust the array facing and tilt seasonally at his discretion.

Figure 9 Flat roof array.

LAS VEGAS, NEVADA
AVERAGE DAILY INSOLATION AVAILABILITY
(KWH/m²)

	JAN	FEB	MAR	APR	MAY	JUN	JUL	AUG	SEP	OCT	NOV	DEC	YEAR
LATITUDE TILT –15(°)													
FIXED ARRAY	4.79	6.09	7.26	8.25	8.38	8.47	7.92	7.91	7.64	6.27	5.05	4.14	6.85
1-AXIS NORTH SOUTH TRACKING ARRAY	6.19	8.21	10.29	11.94	12.47	12.51	11.33	11.11	10.78	8.46	6.58	5.15	9.59
LATITUDE TILT (°)													
FIXED ARRAY	5.64	6.89	7.72	8.22	7.90	7.77	7.35	7.71	7.93	6.92	5.87	4.91	7.07
1-AXIS NORTH SOUTH TRACKING ARRAY	6.86	8.81	10.63	11.95	12.18	12.06	10.96	10.99	11.00	8.94	7.21	5.78	9.78
LATITUDE TILT +15(°)													
FIXED ARRAY	6.16	7.28	7.74	7.73	6.99	6.65	6.40	7.08	7.77	7.17	6.34	5.39	6.89
1-AXIS NORTH SOUTH TRACKING ARRAY	7.27	9.09	10.62	11.60	11.59	11.36	10.35	10.55	10.86	9.10	7.58	6.17	9.68
2-AXIS TRACKING	7.35	9.08	10.64	12.00	12.63	12.81	11.52	11.15	10.98	9.08	7.63	6.27	10.10

LOCATION: 36° 05' N, 115° 10' W, 664 METERS

103F10.EPS

Figure 10 Daily solar insolation data for Las Vegas, Nevada.

2.3.4 Choosing the Panels

Another piece of information needed for the calculations may seem premature at this point—the information from the chosen panel. Since panels are available in fixed specifications, it is easier to pick a particular model, and then calculate the quantity and wiring configuration that would be needed for the application. As mentioned earlier, most designers work with a specific product line. Choose a specific panel capacity from the preferred line of products as a starting point on which to base calculations. If the results of the calculations do not produce an acceptable result for the design and site conditions, another panel capacity can be chosen and the calculations quickly repeated.

The favored panel for this example is a 140W model manufactured by the fictitious TRS Company. The specifications for this panel are shown in *Table 4*.

2.3.5 Array Sizing Calculations

The first calculation in sizing the array results in the array peak amps. This is the amount of current that the array, likely using panels in

Table 4 PV Panel Specifications Example

TRS Model PS140 Panel Specifications (at STC)	
Nominal voltage	12 VDC
Maximum rated power	140W
Maximum power point voltage	18.1V
Maximum power point current	8.34A
Open circuit voltage (V_{oc})	22.2V
Short circuit current (I_{sc})	7.99A
Dimensions	26-1/2" × 60" × 1-1/2" (673 mm × 1524 mm × 38 mm)

parallel, will need to furnish. The equation used is as follows:

$$\text{Array peak amps} = (\text{Average amp-hours per day} \div \text{battery efficiency}) \div \text{peak sun hours/day}$$

The average amp-hours per day were calculated during battery selection at 98 Ah. Battery efficiency is generally assumed as 80 percent, or via a factor of 0.8. Once the battery cannot charge past its 80 percent capacity level, most manufacturers indicate this to be a likely point of replacement. For peak sun hours per day, use the lowest monthly value, with the array faced to latitude. Refer back to *Figure 10*, and examine the table. For Las Vegas, the lowest values are found in the month of December, and the value for a latitude facing is 4.9. Inserting this information yields the following result:

$$\text{Array peak amps} = (98 \div 0.8) \div 4.9$$
$$= 122.5 \div 4.9 = 25$$

Next, calculate the number of panels needed in parallel to develop the targeted array peak amps through the equation:

Number of panels in parallel = array peak amps ÷ maximum power point current

Number of panels in parallel = 25 ÷ 8.34 = 3

Note that the result must be rounded up to the nearest whole number. As is the case with batteries, it is very difficult to install and wire a partial panel.

Next, find the number of panels that must be wired in series to generate the required voltage.

This is based on the nominal panel voltage, not its maximum or open circuit voltage value. The number of panels in series is determined using the following equation:

Number of panels in series = DC system voltage ÷ nominal panel voltage

Number of panels in series = 48 ÷ 12 = 4

With three panels in parallel and four panels in series, the array will require a total of 12 panels. Using the dimensions above, this array will cover roughly 12.3 m² (132.4 ft²). *Figure 11* summarizes the panel and array selection results.

Remember that the specific pole mounting option pictured in *Figure 7* was rated to accommodate up to 65 ft² of array surface area. The array will be arranged into three pole-mounted groups of four panels each. With roughly 45 ft² of array installed on this particular mount, a secure and convenient installation should result.

2.4.0 Charge Controller Selection

Charge controllers have the primary responsibility of delivering energy generated at the solar array to the batteries through the charging process and ensuring that they do not become

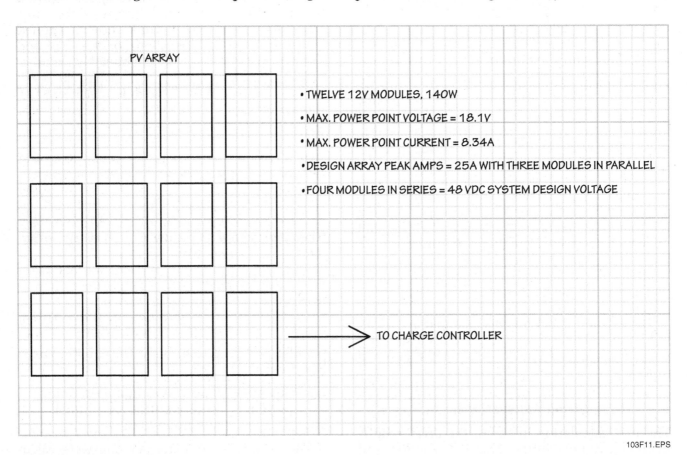

PV ARRAY

• TWELVE 12V MODULES, 140W

• MAX. POWER POINT VOLTAGE = 18.1V

• MAX. POWER POINT CURRENT = 8.34A

• DESIGN ARRAY PEAK AMPS = 25A WITH THREE MODULES IN PARALLEL

• FOUR MODULES IN SERIES = 48 VDC SYSTEM DESIGN VOLTAGE

TO CHARGE CONTROLLER

103F11.EPS

Figure 11   PV array selection results.

overcharged. The different types of charge controllers are reviewed here to facilitate an informed selection. There are several different ways that the controller can fulfill its primary responsibility, and offer other potential features that add value and function to the PV system.

2.4.1 Basic Features

Charge controllers are available in a wide variety of sizes in terms of current capacity. When necessary, if a single controller is unable to handle the current load of the array and battery system alone, two or more controllers can be applied. The array is divided up between the controllers as sub-arrays, with all of the controllers wired back to the same battery bank. Of course, they can also be wired back to separate battery banks as well, but that would create independent or redundant PV systems. In special situations, where an electrical load is both critical and completely dependent on either solar or a hybrid system, redundancy in the charging and power storage functions may be necessary.

Better controllers employ a three-stage charging process. In the initial stage, known as the bulk charge stage, all PV current is delivered to the batteries. As they charge, the voltage at the battery terminals increases. Eventually, the controller senses they have reached the bulk voltage setting, which generally represents 80 to 90 percent of full charge. Once the bulk voltage setting has been reached, the absorption stage begins. As battery voltage continues to rise, the current flow modulates down. This period can be controlled by a timer, or terminated through sensing that battery voltage has reached a second setpoint. Typically this voltage setpoint would indicate near 100-percent full charge. Next, the float stage begins, and the controller begins to maintain battery voltage at yet another setting. The current flow is low and just sufficient to maintain the charge where it is. The float mode can be considered a maintenance or trickle charge. Holding the battery charge current low at this stage, when the batteries are well charged, helps reduce gassing.

A fourth stage, called the equalization stage, is often incorporated into the controller. This stage is used only occasionally, and helps to rejuvenate the batteries. Instead of moving directly to a float mode, the controller continues to charge at the bulk rate past the usual voltage setpoint. The intent is to cause additional gassing and boiling activity in the battery, breaking up and redistributing the electrolyte material. Controllers can be manually placed into an equalization mode or set to initiate the cycle on a timed basis. A typical time period for flooded batteries would be every 30 days. The incidence of equalization should be as set forth by the manufacturer for the specific type of battery being used.

These charging concepts are important to the designer for one particular reason; different types of batteries require different charging values and programs. For example, it is not recommended that gel batteries experience an equalization mode at all. The bulk voltage and float voltage values of the controller should be compatible with the type of battery being used.

Controllers are selected for a single voltage, with the DC voltage of the array, batteries and controller all being equal. The controller specifications will indicate the input and output voltage to be equal. More sophisticated controllers can be field-set for a given voltage for added flexibility in a single product.

Shunt controllers are the simplest type. Their name comes from the fact that they protect the batteries from overcharging by shunting, or shorting, the PV array when the battery charge capacity has been reached. The PV array current then flows through a transistor which converts the excess power to heat. Most controllers, including this simple version, also incorporate a diode in the circuitry to act as a check valve against current flow from the battery bank back toward the array. This type of controller would not be applied in residential PV applications of this size and capacity due to their inefficiency and lack of other desirable features.

For additional capacity with less heat production (shunt controllers can create a tremendous amount of heat when the array is productive), a single-stage controller is the better choice. It switches open the array power circuit, breaking the circuit completely rather than shorting the circuit through a false load. At this level of controller design, and all more sophisticated products, timers are generally built in to top off the battery bank at preselected time intervals, basically entering an intermittent float mode.

Diversion controllers provide another valuable feature to the PV system by making better use of excess energy from the array. Once the battery bank is satisfied, they divert PV array energy to a resistive DC heat load, such as an electric heating element. When the battery bank state of charge is low, all array power goes to the charging system. As the charge capacity increases and approaches full voltage, the controller sends less power (a trickle charge) to the batteries and diverts remaining array power to the resistive load. This energy could be used by a DC heating element added to a water heater, for example. The load should be noncritical in nature, since power to it is not dependable. These controllers may lack current backflow protection, so diodes may need to be installed between the array and the controller.

The most popular type of controller for residential applications is the pulse width-modulated (PWM) controller. This type pulses the charging current to the batteries on and off as they approach a fully charged state during the float mode. The controller detects very tiny changes in battery voltage and sends a short pulse of current to them, possibly several hundred times per minute. This controller is also likely to incorporate a variety of other valuable features and the most flexibility. For all controllers that provide adjustable setpoints of any kind, the designer should take responsibility for specifying the initial setting values.

Other valuable controller features that may be selected by the designer are the following:

- *Low-voltage disconnect (LVD) protection* – Batteries can also be damaged and their life shortened by over-discharging, especially below 20 percent capacity (80 percent discharge). This feature disconnects any existing DC loads from the battery bank when the LVD setpoint is reached. Note that this feature does not disconnect AC loads; the inverter handles this chore. It is capable, however, of starting up an alternate energy source such as a generator.
- *Maximum power point tracking (MPPT)* – In systems without this feature, the voltage of the array is pulled down to something near that of the battery voltage, which is generally 12V to 15V for a 12V battery. This operating characteristic is needed since array power is connected directly to the batteries through the controller, but the batteries would boil over and/or be permanently damaged if higher voltages were applied. However, the array itself may be generating 18V, for example, at its maximum power point. This excess energy is generally shed and wasted in the form of heat at the controller. The MPPT controller tracks the maximum voltage provided by the array and routes array power through an onboard, highly efficient DC power converter to reduce it to the voltage of the batteries. By recovering this usually wasted energy, the MPPT feature can improve the efficiency of the charging system by 10 to 30 percent. In general, the best efficiency boost is in cool, sunny weather when the array voltage is highest, and/or when the batteries have been discharged and their voltage is low. Both of these cases create the greatest diversity between the two voltages and the best opportunity for this feature to pay dividends.

- *Voltage stepdown* – This feature allows a controller to interface between battery banks and arrays of different design voltages. Using higher array voltages decreases wire sizes. In addition, the controller can be installed to serve on a system of lower voltage, and then switched to operate at a higher value for the expansion of the array. This greatly simplifies array expansion if planned in the future, even potentially allowing existing wiring to remain.
- *User interface features* – Items that fall into the category of user interface range from simple indicating lights or meters used to monitor system operation, to digital displays providing multiple operating parameters.
- *Temperature compensation* – Since batteries charge differently at temperatures above and below 77°F (25°C), the charging voltage can be optimized to provide a better quality charge. For example, batteries charging at cooler temperatures may reach their fully charged voltage too soon, before a sufficient amount of current has truly provided a good charge. This feature compensates by adjusting its own setpoint for determining battery charge. The adjustment is made in small increments, such as ±5mV per centigrade degree above or below a battery temperature of 77°F (25°C).

2.4.2 Selecting the Charge Controller

The primary information needed to select a non-MPPT controller includes the following:

- The DC system charging voltage, to be matched to the controller. Note that this is often different than the DC voltage to be routed to the inverter for conversion to AC.
- The single solar panel I_{SC} value. The controller needs to be sized to handle the short-circuit current of the array, with a minimum of 25 percent additional capacity as a safety factor to prevent its overload.
- Number of panels in parallel.
- Maximum DC load values, if any, that the controller will be required to power during a diversion mode.

It is also important to take note of any optional features that the owner may wish to incorporate and add them to the controller selection list. This information may come from the site assessment or through conversation on this specific subject during the design process.

The calculations will provide the maximum array current to which the controller will be exposed. In the stand-alone system example, use the following information to select the controller:

- DC system charging voltage = 12V
- Single panel I_{SC} value = 7.99A
- Three panels in parallel configuration
- No DC load

This system has no need for diversion-style control since there are no DC loads. Also, it will not need a controller that is capable of starting up a backup source of energy such as a generator. However, the client does want pulse width modulation.

The calculation needed to determine the maximum DC solar load current is as follows:

Array short circuit current =
(panel short circuit current × number
of panels in parallel) × 1.25

The 1.25 multiplier simply allows for a 25 percent safety factor, which is common for electrical systems and devices. For this example, the array short circuit current is calculated as follows:

Array short circuit current = (7.99A × 3 panels in parallel) × 1.25 = 23.97A × 1.25 = 29.96A

This is the first specification to be met when selecting the controller—its rated solar current. It should be noted the many controllers can be arranged in parallel to increase the total solar current. For example, two 30A controllers can be arranged in parallel to handle 60A of solar current. If a DC load does exist, the calculation to determine the load current the controller will pass through is:

Maximum load current = DC total connected
load watts ÷ system voltage

Let's review some basic controller specifications for three different controller models from Morningstar Corporation in *Table 5*.

Table 5 ProStar™ Controller Specifications

	ProStar™ Versions		
	PS-15	**PS-30**	**PS15M-48V**
Rated solar current	15A	30A	15A
Rated load current*	15A	30A	15A
System voltage	12/24V	12/24V	48V
Options:			
Digital meter	Yes	Yes	Standard
Positive ground	No	Yes	Yes
Remote temperature sensor	Yes	Yes	Yes

*Low-voltage disconnect included on all ProStar™ controllers.

The Model PS-30 (*Figure 12*) can handle the example 29.96A solar array, and is designed for both 12V and 24V use. Some additional information for this specific model, based on its use on a 12V system, is provided here for review:

- Battery voltage setpoints for sealed (AGM) batteries:
  - Bulk or regulation voltage = 14.15V
  - Float voltage = 13.7V
  - Equalization voltage = 14.9V
- Disconnects load from batteries at voltage = 11.4V
- Reconnects load to batteries when voltage = 12.6V
- PWM battery charging
- Incorporates all four stages of battery charging
- Three-position battery select: gel, sealed, or flooded
- Parallel operation for up to 300A solar current
- Temperature compensation, with remote temperature probe to monitor battery temperatures
- Current-compensated LVD for DC loads
- LED's indicate battery status and faults
- Capable of 25 percent overloads
- Remote battery voltage sense terminals
- Digital meter
  - Highly accurate voltage and current display
  - Low self-consumption (1 milliamp)
  - Includes manual disconnect button
  - Displays five different protection functions and disconnect conditions
  - Self-diagnostics (self-test)

This controller represents a reasonable choice for this system. It does not provide any additional capacity for expansion, but a second controller can be added in parallel if expansion is needed in the future. The controller represents a small cost in the total installation, compared to many other components. *Figure 13* shows the selection results.

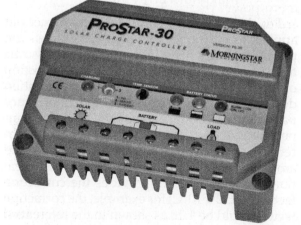

103F12.EPS

Figure 12 Morningstar ProStar PS-30 controller.

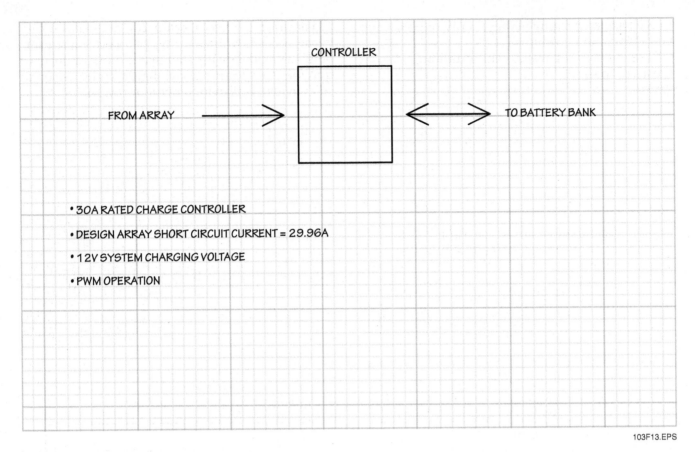

CONTROLLER

FROM ARRAY

TO BATTERY BANK

- 30A RATED CHARGE CONTROLLER

- DESIGN ARRAY SHORT CIRCUIT CURRENT = 29.96A

- 12V SYSTEM CHARGING VOLTAGE

- PWM OPERATION

103F13.EPS

Figure 13   Controller selection results.

2.4.3 MPPT Controllers

When the valuable MPPT feature is desired in the controller, some additional issues must be considered.

- A more precise voltage calculation is necessary to ensure that the controller can handle the maximum voltage produced by the array:

Selected panel V_{oc} × number of panels in series × temperature correction factor = maximum array voltage

Per *NEC Article 690.7(A)*, this procedure applies to all mono- or polycrystalline silicon panels, and the resulting voltage shall also be the value used as the rating in the selection of cables, disconnects, overcurrent protection devices, and other similar components. Since panels produce higher voltages at lower temperatures, the temperature correction factor corrects the STC voltage for the lowest expected ambient temperature of the installation location. *NEC Table 690.7* provides the correction factors. At 32°F (0°C) for example, the correction factor would be 1.10 as shown in the referenced table. If the manufacturer does provide open circuit voltage temperature coefficients in their specifications, those values are directed for use in lieu of *NEC Table 690.7*. Once the maximum array voltage is determined from the equation above, it must be compared to the controller specifications to ensure that it falls within the controller's acceptable voltage range.

- The maximum output amperage of the controller needs to be higher than the maximum input amperage produced by the array. If this is not the case, and the output amperage of the array is equal to or greater than the output amperage of the controller, the controller will not be able to make use of the additional capacity MPPT provides.

MPPT is a feature that can also be applied to inverters that are capable of charge control, and to inverters used in grid-tied systems without batteries. This latter application of MPPT will be discussed later in this module.

2.5.0 Inverters

It is well known that most common electrical loads are designed to utilize alternating current (AC) in lieu of direct current (DC). The utility grid provides AC power. Unfortunately, PV systems, as well as batteries, are capable only of providing and storing DC power.

The inverter provides the interface between the two types of power, regardless of whether the system is a stand-alone or grid-tied application. Inverters have historically been one of the weak spots in PV systems due to inefficiencies and losses in the conversion process. More recent inverters, employing better and newer technology, are more efficient and capable of producing cleaner power. In fact, new records were recently set for inverter efficiency levels by a unit which achieved over 99 percent efficiency. Inverters which produce a sine wave are the primary type used in PV systems today, and will be the only type considered here.

When selecting an inverter, one consistently important factor is efficiency. Most inverters can reach efficiency levels of 90 percent or better, but this is not the case under all operating conditions.

Inverters must be selected for a load capacity large enough to handle the full load that can be applied, but in reality they spend most of the time passing less than their design current capacity. The inverter efficiency curve in *Figure 14* shows that the efficiency of this unit is quite good for the vast majority of the AC wattage load range. As the load is significantly reduced however, the inverter's ability to efficiently convert DC power to AC drops off substantially. An understanding of how each system will be operated should be factored into the inverter selection process. If it is understood that the system will operate with very small AC loads most of the time, then careful scrutiny of the efficiency curve will be in order to pick the best possible inverter for the application.

It is important that all inverters exhibit good characteristics in the control of harmonic distortion and frequency (60 Hz for the U.S.). They should also be easy to service and reliable. Some

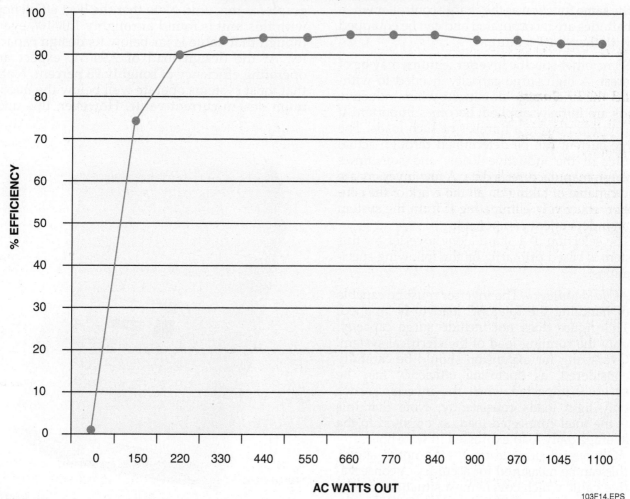

103F14.EPS

Figure 14   Inverter efficiency curve.

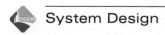 System Design

other features that are available and may be needed for a given installation are as follows:

- *Series or parallel operation* – In complex systems, using inverters in series or parallel allows a system to serve AC systems of higher voltage or higher load current.
- *LVD operation* – This provides a warning and/or shutoff of power to the AC loads if the batteries become discharged down to the LVD setting.
- *Ventilated or sealed enclosures* – Both are available, depending on the environment. Sealed units prevent water, dust, and bugs from entering.
- *Hybrid operation* – Since certain systems incorporate a generator which provides an alternate source of AC power, some inverters can control the operation of that source to allow for battery charging when PV power is unavailable.
- *Remote control or data logging* – Some inverters are especially suited for remote installations and those that are not closely monitored by providing for remote access and logging of operating conditions.
- *Charge control with maximum power point tracking (MPPT)* – Inverters are available that provide the same function as the charge controller when batteries are incorporated and can be equipped with the MPPT feature.

Some other specific inverter features may be of interest. A high surge capacity, needed to withstand the brief surge in current when inductive loads are initially applied, becomes important if the system powers a number of such loads. The surge current can be determined through actual testing of the inductive loads, and sometimes through manufacturer's data. A few inverters are also capable of taking on all the work of the controller, effectively eliminating it from the system as an independent component.

The selection of a stand-alone inverter for this system is based primarily on the following specifications:

- *AC load wattage* – The inverter must be capable of handling the total AC load to be applied. This figure does not include surge capacity, only the running load of the electrical system. Oversizing for expansion should be carefully considered, as operating efficiency may be severely impacted when the inverter carries only light loads consistently. Note that this is the total connected load, as opposed to the average daily load expressed in watt-hours.
- *AC projected surge wattage* – As noted above, this can be measured for accuracy. A conservative value which covers most situations can be determined by multiplying the AC load wattage by 3.

- *AC output voltage* – Inverters can be selected to provide 120 VAC, 208 VAC, or 240 VAC, depending upon the design of the system.
- *DC input voltage* – The inverter must be selected at the proper DC system voltage to be applied.
- *Frequency* – For the US, only 60 Hz models may be considered.

This information has already been determined from previous calculations for the example system. The selection process begins with the following information:

- AC load wattage = 2,940W
- AC projected surge wattage = 2,940W × 3 = 8,820W
- AC output voltage = All connected loads operate at 120VAC
- DC input voltage = 48V
- Frequency = For the US, it will always be 60 Hz

One possible selection is this inverter, the Sunny Island 4248U (*Figure 15*) from SMA America. This unit is nominally rated for 4,200W of power at 77°F (25°C), and is designed for a 48VDC input. *Figure 16* shows the efficiency graph for this unit. Note that the best efficiency with this unit is found at roughly 1,000W, even though that value is far below its design capacity. At the design load of 2,940W, expect an operating efficiency of roughly 93 percent. Note that most systems operate well below the maximum design current value. However, this unit

103F15.EPS

Figure 15   SMA Sunny Island 4248U inverter.

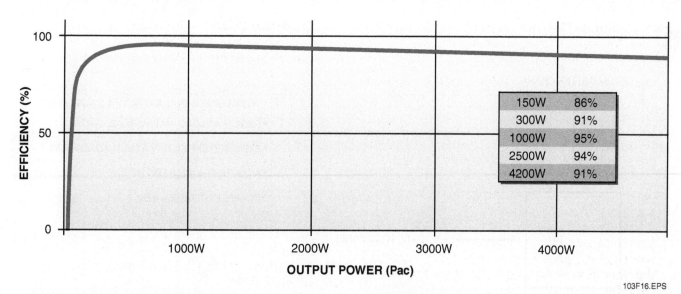

150W	86%
300W	91%
1000W	95%
2500W	94%
4200W	91%

103F16.EPS

Figure 16   Sunny Island 4248U efficiency curve.

operates efficiently down to just a few hundred watts of power.

Typical inverter specifications include the following:

- *Continuous AC output at 77°F/113°F = 4,200/ 3,400W* – This specification conveys an important piece of information: that temperature of the inverter and its environment is a factor in capacity. Note that a 36°F increase in temperature reduces the available power from an inverter significantly.
- *Continuous AC output at 77°F (25°C) for 30/5/1 min = 5,400/6,200/11,900W* – This identifies the inverter's ability to withstand overloads and surges, and the time periods it can be expected to survive these load values.
- *Maximum current (peak value) for 100 ms = 100A* – For a very short period of time (1/10 of a second, to be precise), this inverter can handle as much as a 100A load.
- *Battery voltage (range) = 48V (41 – 63V)* – Inverters can typically handle DC input voltages within a significant range of values.
- *Weight: 39 kg* – This value is shown here only to provide an idea of the unit's weight. Inverters can be heavy.

Now that an inverter has been selected, all important specifications and the primary equipment for this stand-alone project have been identified (*Figure 17*).

2.6.0 System Design and Equipment Review

The calculations and equipment selections include the following:

- Total AC Average Load Daily = 4,239 watt/ hours
- Total AC load wattage (peak load) = 2,940W
- AC voltage = 120V
- Design DC voltage to the inverter = 48V
- Array peak amps = 25A
- Design days of autonomy = 4
- Batteries required = (16) 12V, 212Ah batteries, Model B-2120. Four sets of four paralleled batteries are connected in series.
- Solar array = (12) TRS Model PS140 panels, 12V, 140W. Three sets of four paralleled panels are connected in series.
- Array short circuit current = 29.96A
- Controller = ProStar PS-30, rated at 30A load capacity and solar array capacity, for 12V battery charging
- Inverter = SMA Sunny Island 4248U, rated for 4,200W continuous output at 77°F

Since the next step in the design process is to size the interconnecting power wiring, both AC and DC, all pertinent data from the manufacturers of these components must be acquired before proceeding. With this information available, the next step is wire sizing.

3.0.0 System Wiring

Properly sized and installed wiring is essential to the function of any PV system. Beyond the wiring itself, other devices such as overcurrent protection devices, power disconnects, and surge suppressors must be considered and included in the electrical layout. Since the utility grid is based on AC power, the DC side of the PV system may seem a bit foreign to the designer, further complicating the chore.

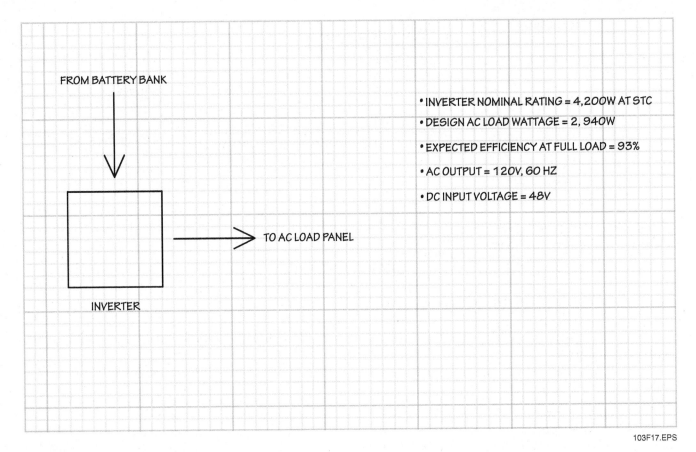

FROM BATTERY BANK

TO AC LOAD PANEL

INVERTER

- INVERTER NOMINAL RATING = 4,200W AT STC
- DESIGN AC LOAD WATTAGE = 2,940W
- EXPECTED EFFICIENCY AT FULL LOAD = 93%
- AC OUTPUT = 120V, 60 HZ
- DC INPUT VOLTAGE = 48V

103F17.EPS

Figure 17 Inverter selection results.

3.1.0 Wire, Cable, and Raceway

The color conventions for DC systems are red and black. Use colored sleeves or phase tape to identify cables where necessary. Some common wire types used in PV systems are THHN for dry, indoor applications, and TW, THW, and THWN for both indoor and wet outdoor locations in conduit. Wire used in exposed applications outdoors must be marked as sunlight resistant. A comprehensive listing of conductors and their individual specifications can be found in *NEC Table 310.104(A)*.

Underground feeder (UF) and underground service entrance (USE-2) are cable grades that are typically used in PV system installations.

Both cables and raceway, with individual conductors pulled, are used on solar projects depending upon the duty and code requirements.

Wire sizing for solar PV systems is not much different than any other type of wire sizing. First, the amount of current that the wire needs to carry is calculated, and then a factor of 1.25 is applied because PV is considered a continuous duty load. Of course, the voltage must also be known to ensure the selected wire has a sufficient insulation value. The *NEC*® specifies the current-carrying capacity of wire in a variety of tables.

3.2.0 Wiring Diagram

Figure 18 shows one example of a basic, stand-alone PV system wiring diagram. Obviously each diagram and system differs in a variety of ways while still being functionally similar. As mentioned previously, designers generally align themselves with a particular line of products and utilize those products in most of their projects. This allows them to become very familiar with the specific wiring requirements of each component. It is imperative that the designer become intimately familiar with the wiring layout for the system. When there is any doubt or discrepancy at the field level during the planning and installation process, the designer will likely be called upon to assist and ensure that the wiring is done correctly.

Beyond the wiring which carries power, there can also be a significant number of control wires involved with an installation. When components are wired for off-site monitoring, for example, communication cables are in order. The final design may also incorporate system indicators which are remote from the installation itself to allow the owner to monitor performance without physically accessing the installation.

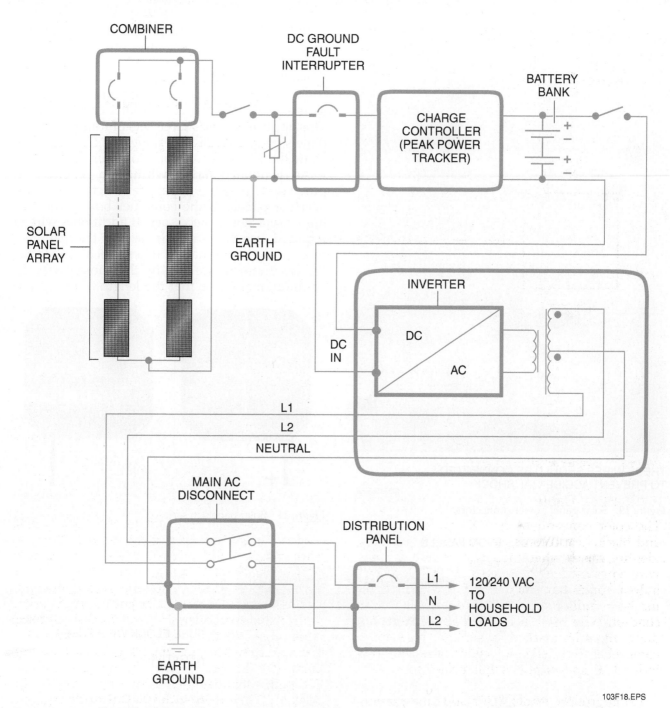

Figure 18 Typical stand-alone system diagram.

103F18.EPS

3.3.0 Solar Array Wiring

Solar panels generally come with wire leads and connectors for easy installation. These leads, such as the ones shown in *Figure 19*, often utilize a special type of connector. The connectors pictured here conform to the terminal specification RWN-2. Cables with mating terminals can either be purchased or field-fabricated with the proper tool to secure the terminal to the wire.

Wiring from individual solar panels is normally connected at a solar combiner (*Figure 20*).

The combiner, in its simplest form, is little more than a junction box with one or more terminal strips. The specific model shown here also incorporates a power disconnect. This optional safety feature provides a means of isolating the power output of the array from the rest of the system for service, eliminating the need for a separate power disconnect to be installed between the array and the charge controller or inverter. On larger projects, multiple combiners may be used, with the DC output from each of them returning to a central combiner, also referred to as a re-combiner.

For this project, the power leads must be brought from each panel back to the combiner and wired in the proper series and parallel configuration. This application uses three sets of four paralleled panels connected in series. *Figures 21, 22, and 23* show the wiring for various configurations.

As mentioned earlier, wiring can generally be purchased from the panel manufacturer to connect the panels to the combiner box, so no sizing or selection is necessary. If additional wire is required, use the same wire size as the manufacturer, and ensure that all other specifications for the environment are met by the chosen wire. In addition, make sure that the length of the run to

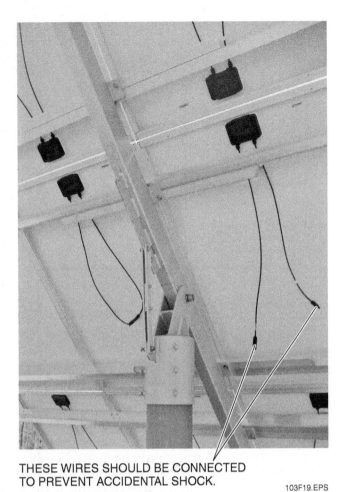

THESE WIRES SHOULD BE CONNECTED
TO PREVENT ACCIDENTAL SHOCK.

103F19.EPS

Figure 19   Solar panel power connectors.

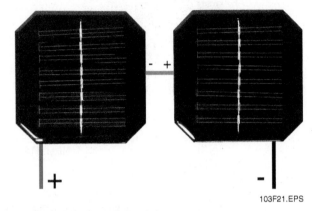

103F21.EPS

Figure 21   Solar panels in series.

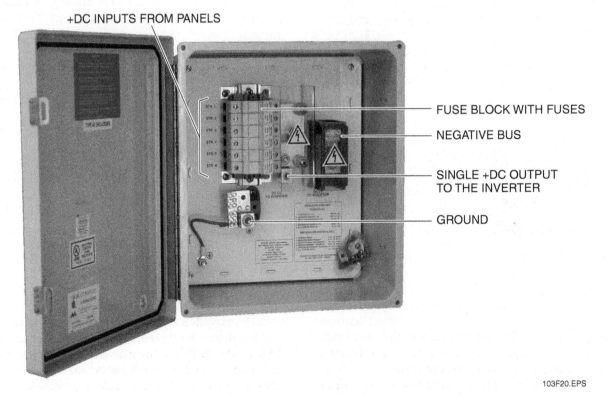

+DC INPUTS FROM PANELS

FUSE BLOCK WITH FUSES

NEGATIVE BUS

SINGLE +DC OUTPUT
TO THE INVERTER

GROUND

103F20.EPS

Figure 20   Solar combiner with disconnect.

NCCER — *Contren® Learning Series*   57103-11

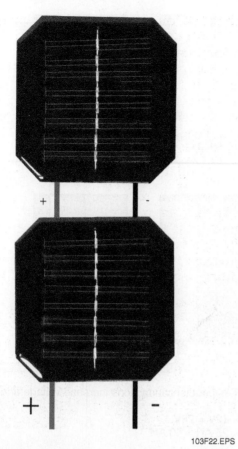

Figure 22 Solar panels in parallel.

103F22.EPS

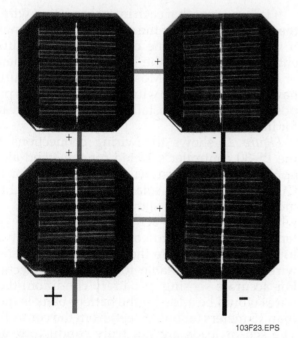

103F23.EPS

Figure 23 Solar panels in series-parallel configuration.

the combiner is not excessive for the wire size. #12 AWG is a typical wire size for PV panels and is used on the 140W panels selected for this system.

Begin the wire sizing calculations by choosing the appropriate wire size to connect the array

combiner box to the charge controller. The highest potential current flow through these conductors is the array short circuit amperage of 29.96A. Multiplying by 1.25 to determine the ampacity for wire sizing provides a value of 37.45A. This is typical for wire sizing to ensure that conductors do not ordinarily carry more than 80 percent of their rated capacity. However, per *NEC Sections 690.8(A)(1) and (B)(1),* this result must be multiplied by 1.25 a second time to compensate for potential excess current that can occur during periods of extreme solar insolation, such as that caused by bright reflections on the panel. This is a wire sizing practice unique to solar applications. The calculation can be done in a single step by multiplying the original value by a factor of 1.56. Applying these rules to the calculations results in a wire size of 46.73A. Per *NEC Table 310.15(B)(16)* for conductors in conduit, raceway, or directly buried in earth, #6 AWG THW wire is adequate when using cable or conduit. Note that *NEC Table 310.15(B)(16)* covers only applications with no more than three current-carrying conductors. See *NEC Table 310.15(B)(2)(b)* for applicable correction factors.

Since this wiring connects the outdoor array combiner with the indoor controller, it must be run in metal raceway. Metal raceway is generally easier to seal at wall or roof penetrations than cable, and offers better protection and longevity when exposed to weather and sunlight. If there are any plans for future expansion of the array, the raceway can be sized a bit more generously to allow for larger wires to be pulled in later. Selection of a larger raceway will not impact present wire sizing.

In addition to current flow, a designer must also consider voltage drop to minimize significant losses. Voltage drop in a PV system should be limited to no more than 3 percent. Sizing for voltage drops closer to 2 percent on the DC side is recommended for all PV systems.

Voltage drop can be calculated manually or by using a voltage drop calculator available via the internet. The *NEC*® does not provide tables for voltage drop calculations. Regardless of the chosen method, the following information is required:

- Conductor material (copper or aluminum)
- Voltage
- Current load
- Length of run

This example uses a conductor distance from the combiner to the charge controller of 33 feet. Using one simple online voltage drop calculator, it is shown that #6 AWG wire reaches a 2.8 percent voltage drop when carrying a 47A current load at a 104°F (40°C) conductor temperature.

Running the same calculation based on a 167°F (75°C) conductor temperature results in a 3.2 percent voltage drop. Many such calculators assume the 167°F (75°C) conductor temperature, as this is the temperature on which *NEC Chapter 9, Table 8, Conductor Properties* bases wire resistivity. More complex calculators like this one can accommodate other unique operating conditions.

Since the 3.2 percent voltage drop is greater than desired, the wire size is increased. Recalculating based on #1 AWG wire results in a 1 percent voltage drop at 33 feet. Keep in mind that voltage drop values in wire are proportional to the voltage, such that wire carrying 24V travels only half the distance before experiencing a 2 percent voltage drop than it does at 48V. This is another example of why higher voltages on the DC side can be an advantage in solar PV system design. The final selection will be #1 AWG wire for this application, to accommodate the wire run to the controller.

3.4.0 Battery and Controller Wiring

Batteries come with several different styles of terminal connections, and in some cases they can be factory-ordered with a specific terminal type. Many PV system batteries also have four terminals in lieu of two, to accommodate the extra connections that are often needed and to simplify wiring in various configurations. Using two connections for each leg of power provides redundancy and splits the current load of the terminals in high-current applications. Using all four terminals is therefore generally optional, but they should always be used in applications which exceed 400A.

One example of series-parallel wiring of batteries is shown in *Figure 24*. This shows two batteries in parallel and two in series.

In a larger battery bank, a few wires will carry more current than the series-parallel wiring from battery-to-battery. Consider *Figure 25*. Note that the wires marked C will carry the maximum amount of current to the batteries. Wire B will carry 2/3 of the current as C, and wires marked A will carry 1/3 of the current as C. As a result, wires B and A can be sized at a different amperage and be a smaller gauge, although it is not a necessity. This approach to wiring the battery bank is ordinarily used in applications that require relatively low power flow.

Figure 26 provides an example of a battery bank wired in a series-parallel configuration that uses busbars to collect the power into a single, noninsulated conductor. Note that the wiring here carries a balanced load, as opposed to the configuration shown in *Figure 25*.

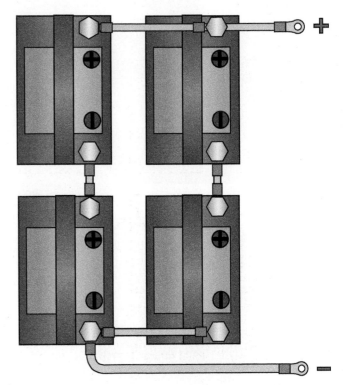

Series/parallel connections increase voltage and capacity:

2 × 12V = 24V
2 × 25 AH = 50 AH

103F24.EPS

Figure 24 Series-parallel battery connections.

It is important that each interconnecting wire in the battery bank be of the same gauge and length. If the resistance in one or more wires is different than the others, the batteries do not charge or discharge equally. Most battery manufacturers have accessory sets of wires premade for this purpose, with factory-installed terminations ready for installation.

Figure 27 shows the wiring connections for the PS-30 controller selected for the stand-alone application. Two terminals are provided for the DC input from the solar array. Several LED's indicate the present status of the batteries and whether charging is active. There are also two terminals provided to connect the battery bank to the controller. Note the two additional sensor wires that are available for battery connection. For accurate sensing of battery condition, these wires would be added if the battery bank is more than 15 meters (about 46 feet) from the controller. The sensor leads are not truly conductors, and thus can be small gauge wire such as #16 AWG.

In most all cases, the wiring between the PV array and the controller can be the same size as the wiring from the controller to the battery bank, since the controller simply passes the current through. However, if the wire has been upgraded significantly due to the length of the run and the

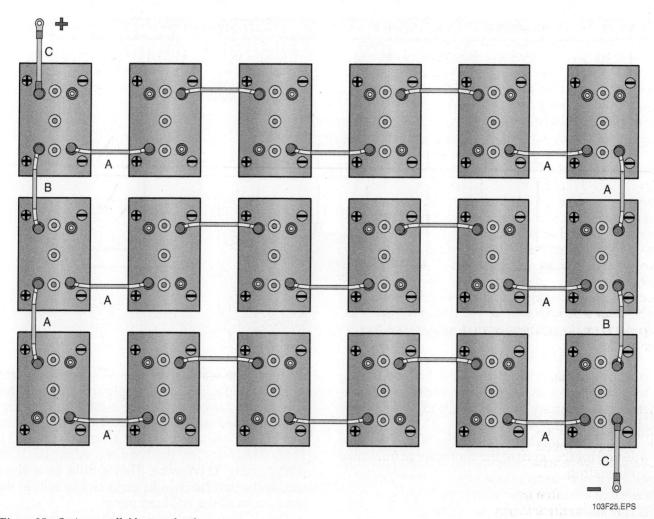

Figure 25   Series-parallel battery bank.

103F25.EPS

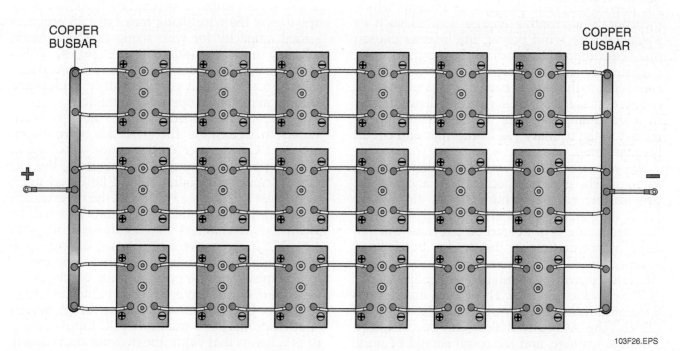

Figure 26   Series-parallel battery bank with busbars.

103F26.EPS

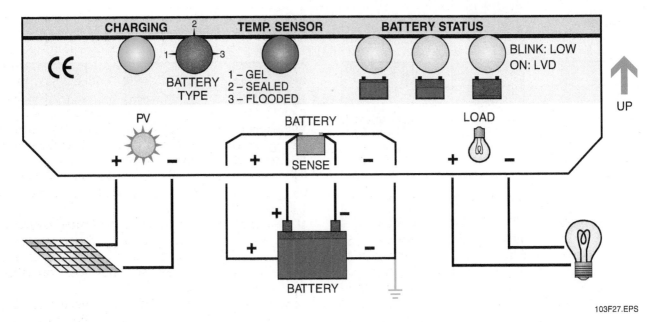

Figure 27 ProStar PS-30 controller wiring connections.

resulting voltage drop, then the battery wiring may be resized, as the distance between the controller and battery bank is likely to be relatively short.

For this project, maintain the #1 AWG size selected for the input to the controller to minimize voltage drop.

3.5.0 Inverter Wiring

The power wiring for the chosen inverter includes the DC power supply from the batteries and the AC power supply to the load. In systems which incorporate alternative energy sources such as a generator or wind power, the inverter chosen must be equipped to handle these inputs as well.

Generally, the inverter's DC power input is connected to the battery bank at the same point of connection as the controller and DC loads, if any. In a stand-alone system, the inverter output is routed to the primary AC load center and circuit breaker panel.

AC wire sizing for inverters is rather straightforward. The inverter specifications and manual provide the necessary information, along with *NEC®* wire sizing tables. The current flow used for wire sizing is not based on previous calculations of power flow in the design, but instead is based on the inverter's maximum continuous AC rated power at standard test conditions per *NEC Section 690.8(A)(3)*. The chosen inverter has a maximum continuous output of 4,200W at 77°F (25°C). The AC current flow can be calculated from the wattage, and the result should be quite close to that noted in the inverter specifications.

Dividing the wattage by the AC supply voltage of 120V yields 35A. Next, apply the ampacity for wire sizing factor of 1.25 for a total ampacity of 44A (rounded up to the nearest amp). The wire size is then calculated as are all others. Per *NEC Table 310.15(B)(16)*, the required wire size is #8 AWG, using THW wire. This should be a short run, as the inverter would generally be within the same room as the AC load panel.

The DC input can be calculated in the same manner. Use the same output wattage value of 4,200W of AC power, and then divide by the DC input voltage of 48V to determine a current flow of 87.5A. Multiplying by the wire sizing factor of 1.25 provides a total ampacity for wire sizing of 110A. Again referring to *NEC Table 310.15(B)(16)*, #2 AWG THW wire could be used. Although this procedure sounds reasonable, it could result in undersized wire for all potential operating conditions.

Other inverter specifications must be consulted to determine the basis for wire sizing on the DC side. Per *NEC Section 690.8(A)(4)*, which applies to stand-alone inverters, the DC power input wire sizing should be based upon the lowest DC voltage value that the inverter can accept and still maintain operation. This will be a value somewhat less than the design voltage of 48 VDC. It can make a significant difference in the calculations for wire sizing, but is necessary to accommodate the possibility of power being delivered to the inverter at a value less than 48V. The Sunny Island 4248U inverter, per its specifications, has a minimum DC input voltage of 41V. Below that value, the inverter shuts down. The calculation then, including application of the ampacity factor, results in an ampacity for wire sizing of 128A, not 110A. This changes the wire size from #2 AWG to #1 AWG per *NEC Table 310.15(B)(16)*. Since this is on the DC side, check the voltage drop using the #1 AWG wire. It is a very acceptable 1 percent at 10 feet.

A summary of the wire sizing for this example is displayed in *Figure 28*.

3.6.0 Overcurrent Protection and Disconnects

Essentially the *NEC*® calls for some means of overcurrent protection for all current-carrying circuits to prevent a continuance of current flow which exceeds the wire's or device's ampacity. Disconnect switches are also very important in system design. Per *NEC Section 690.15*, a means must be provided to disconnect equipment, such as inverters, batteries, charge controllers, and the like, from all ungrounded conductors of all sources. In many cases, the overcurrent protection device also acts as the means of disconnect for service, since the code does allow for this. The means of disconnect must be permanently marked to identify it as such.

3.6.1 Array

The *NEC*® requires that solar arrays that operate as a system rated at 50V or less can be considered a single source of power and incorporate a single overcurrent protection device for the complete array. When arrays are designed to operate in excess of 50V, individual panels may require fusing in addition to the overcurrent protective device for the full array.

The wire sizing calculations for the solar array connections to the charge controller resulted in a current value of 46.73A. *NEC Section 240.4(B)(2)* allows selection of the next highest rated overcurrent protective device. Therefore, the overcurrent protective device for the array wiring into the controller will be a 50A circuit breaker rated for use in a DC system.

Remember that #1 AWG was selected for the array power wiring to the controller to reduce voltage drop. In this case then, the #1 AWG wire can handle a higher current than necessary. This upgraded selection in wire size also affords some room on sizing overcurrent protection and adds an additional electrical safety factor to the design. However, it is also important to note that financial considerations must be a significant part of a

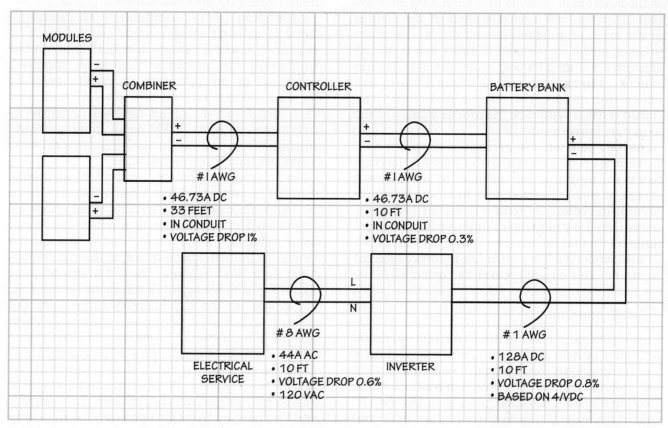

103F28.EPS

Figure 28 Component wire sizing summary.

 System Design

designer's thought process, so needless oversizing of wire must be avoided.

A means of disconnect is also in order for the array wiring to the controller. The combiner box selection incorporated a means of disconnect, but not overcurrent protection. Per the details of *NEC Section 690.17*, an overcurrent protective device can also function as the required means of disconnect. As a result, the designer may choose to use a simple combiner and apply a separate circuit breaker or fusible disconnect adjacent to it.

3.6.2 Charge Controller

The DC power wiring from the charge controller to the battery bank must also be protected by an overcurrent protective device. The sizing process is basically the same—base the selection on the ampacity used for wire sizing, and when the ampacity does not match a standard size, select the next higher available overcurrent protective device. Since the output and input wiring have the same ampacity, a 50A circuit breaker will be used.

3.6.3 Inverter

The AC output power wiring from the inverter to the load panel must be protected with an overcurrent protective device located at the output of the inverter. In stand-alone systems, the main breaker of the load panel is sized based on the ampacity value resulting from the wire sizing exercise. The example system based the wire size on a value of 44A. Therefore, a 45A single pole circuit breaker would be the proper choice for the load panel inverter input.

As is the case with the array, *NEC Section 690.15* permits a single means of disconnect for the combined AC output of one or more inverters. Since inverters generally provide a significant amount of built-in protection against overload and incorporate sensitive electronics into this function, it is quite likely that the inverter itself will shut down or disconnect itself from the load when problems occur, before a circuit breaker has responded.

3.6.4 Grounding

Proper grounding is essential to the safe operation of the PV system and must be considered during the design phase. Per the *NEC®*, when PV system DC voltage exceeds 50V, then one conductor of a two-wire system must be grounded. On the DC side of the system, the negative conductor is chosen to be grounded, and locating it close to the array is advised to help the system survive powerful surges associated with lightning. However, *NEC Section 690.35* allows PV power systems to operate with ungrounded photovoltaic source and output circuits where the system complies with *NEC Sections 690.35(A) through 690.35(G)*. These items outline the necessary system characteristics to install and operate an ungrounded system. These requirements include, but are not limited to the following:

- All system disconnects must be in place as required under *NEC Article 690*.
- All overcurrent protective devices must be installed per *NEC Section 690.9*. Both conductors must incorporate an overcurrent protective device.
- Ground fault protection must be provided on all PV system source and output circuits.
- Specific warning labels must be applied to all junction boxes, combiner boxes, disconnects, and devices where energized, ungrounded circuits may be exposed during servicing.

On a system where grounding of the conductors is planned, it is important that it be grounded at only one point, per *NEC Section 690.42*. If grounding connections are made at more than one point, the potential exists for the grounding conductor to carry a portion of the current normally carried by the conductor.

Grounding of the conductors as discussed above is referred to as system grounding. The issue of grounding though, does not stop with the conductors. Equipment grounding is required regardless of the approach to system grounding, and system voltage is irrelevant here. Per *NEC Section 690.43(A)*, all exposed noncurrent-carrying metal parts of PV panel frames, electrical equipment, and conductor enclosures must be grounded regardless of voltage.

Per *NEC Section 690.4(C)*, individual solar panels must be connected in a manner which allows the removal of one panel without interruption of a grounded conductor from another PV source circuit. That is a fairly simple matter that can be handled at the combiner box during the installation.

Consult *NEC Section 690.45* for information regarding the sizing of equipment grounding conductors, as the requirements differ with the use and application of ground fault protection devices. Ground fault protection devices monitor the current of conductors and neutral wires. Should an imbalance occur, it is a signal to the device that current is flowing somewhere it should not, presumably to ground. This current could be passing through a person who has come into contact with a live conductor. The device then disconnects the current-carrying conductor from the circuit, interrupting the flow of power, and typically provides a visual indication that ground fault protection has been enabled.

Ground fault protection is generally required on grounded and ungrounded PV arrays, but there are some exceptions noted. Consult *NEC Section 690.5* for details regarding PV systems and ground fault protection. It is not uncommon for the system inverter to incorporate ground fault protection, eliminating the need for a separate, stand-alone device.

The sizing of equipment grounding conductors is covered in *NEC Section 250.122*. When a system is equipped with ground fault protection, the grounding conductor can be as large as the conductors themselves, but it is never required to be larger. *NEC Table 250.122* shows the minimum size the grounding conductor can be, and it is based upon the sizing of the overcurrent protection. However, it is important to note that when current-carrying conductors have been oversized to allow for reduced voltage drop (the most likely reason for oversizing), the grounding conductor must be increased in size proportionately, per *NEC Section 250.122(B)*.

When ground fault protection is not present in the system, wire sizing for the equipment grounding conductor is then based on 125 percent of the PV array short circuit current. The *NEC®* requirement regarding conductors which have been oversized applies here as well—the grounding conductor size must be proportionately increased.

Since equipment grounding is always required regardless of the approach to system grounding, all PV system equipment grounds must eventually terminate at a grounding electrode, literally installed in the ground. The electrode will be connected to the system through the grounding electrode conductor. The AC and DC circuits of the PV system can share a grounding electrode, or they can be separated for convenience sake. When two separate grounding electrodes are utilized though, they must be bonded together. Consult *NEC Section 690.47(C)(2)* for additional details regarding grounding electrode arrangements, and *NEC Section 250.166* for the sizing of grounding electrode conductors.

Some relatively new provisions found in *NEC Sections 690.48 and 690.49* require that jumpers be present for equipment and system grounding conductors to ensure continuity of the grounding circuit in the event an inverter is removed for service or replacement. Although it is unlikely a hazard would be immediately created without such jumpers in place, they are required as a precaution against injury for a worker who may not be aware the grounding circuit is no longer intact.

3.6.5 Surge Suppression

Mounting solar arrays high and in wide open areas attracts lightning and the resultant power surge it creates to the metal-framed panels. PV system designers and installers need to plan for and provide appropriate surge suppression protection.

The *NEC®* has little to say regarding surge suppression devices, since their use is more about equipment protection than safety and therefore left to the discretion of the owner or designer. Surge suppression devices in general are designed to shunt transient voltages from the equipment and conductors to the earth grounding system. PV system designers must consider the risk versus cost and determine the level of protection that should be employed.

Surge protection can be applied on both the AC and DC sides of the system, depending on the location of the perceived threat. The threat of surge is generally no greater on the AC side in PV systems than in any other grid-connected electrical system. As noted above, the array is the location of greatest exposure to lightning-induced surge, and represents the most likely point to begin with surge suppression application.

Since the array has the greatest potential to experience surges, the combiner box is an appropriate location to apply some protection, especially in stand-alone systems. *Figure 29* shows one type of surge suppression device. These devices are first selected based on the system operating voltage. The next specification of concern is related to the surge current capacity. A 40,000A value would be considered common and generally sufficient, while a unit rated at 100,000A would be considered

SURGE SUPPRESSOR

103F29.EPS

Figure 29   Surge suppression device.

heavy duty to minimize the risk of even further damage.

Wiring the surge suppression devices is straightforward. There are basically three wiring connections. The unit is simply wired in parallel with the two DC conductors, and the third connection goes to the earth grounding system.

4.0.0 GRID-TIED SYSTEMS

Grid-tied systems are installed in conjunction with utility-provided power. As a result, they can be designed to provide all the power needed for a structure, or only a fraction thereof. In addition, they can be designed to provide even more power than needed, with the excess power generated being transmitted upstream to the utility grid.

Grid-tied systems have a number of advantages. Each of the approaches to grid-tied system design have some disadvantages as well, which must be discussed in depth with the user. One critical aspect to discuss is the need or desire for an uninterruptible power supply. Many people simply assume that solar systems allow the owner to have power at any time when the power from the utility has been interrupted for some reason. However, this is certainly not the case. When grid-tied systems are installed without a battery bank for power storage, the array is disconnected completely from the grid, as well as the load, when utility power is interrupted. This must be done for safety purposes, preventing line workers or others servicing the grid from being electrocuted by what they believe to be a circuit without power. When battery banks are employed, the array is disconnected from the grid but the battery system remains connected to the load through inverters designed specifically for this purpose.

Before initiating the design process, this issue and a number of other functional characteristics must be discussed to ensure that the final product meets the needs of the owner. Grid-tied systems without batteries are fairly simple in design, incorporate fewer components, and are far less expensive overall. Not only is the initial cost of a system with batteries much higher, but the life cycle cost continues to grow due to added maintenance and repair of the additional equipment.

In some applications, the need for an uninterruptible power supply cannot be avoided. However, when it is avoidable, the simplicity and reduced installation cost of the grid-tied system without batteries involved is very attractive. Future expansion of the system is greatly simplified, due to the modular nature of PV arrays. Inverters can be oversized in the beginning (with some loss in efficiency), or they can be added as the system grows by operating them in parallel. This feature allows an owner to invest in a smaller system in the beginning, adding capacity as needed at a later date.

4.1.0 Grid-Tied System Component Selection

The *Site Assessment* module used an average home in South Carolina as an example project. Some of the information generated from it will be used in this design example. More complex issues in sizing calculations will also be explored. Forms have been provided for use in the grid-tied component selection process and can be found in *Appendix B*. See *Figure 30* for a simple component form for a grid-tied system.

This example will be a simple grid-tied system with no battery storage. Although battery storage would be a distinct advantage to the homeowner in the event of a blackout, it was determined to be cost-prohibitive at this juncture. The best opportunity in this case is to invest well in the power production capabilities and consider future expansion and battery storage if autonomy becomes more important.

In the site assessment example, it was determined through energy bills that the residence utilized an average of 54kWh per day. It was also expressed that the owner wished to generate 50 percent of the average daily electrical needs at this time. It follows then, that the target would be 27kWh per day. Since there are no batteries, no charge controller needs to be selected. This example focuses on the sizing and selection of the panels and the inverter used in grid-tied systems.

4.1.1 Sizing the Array

The first value needed in the array calculations is the average peak sun hours each day for this location, which is the small town of Pelzer, SC. The nearest city with readily available data in the National Renewable Energy Lab (NREL) tables is Greenville, SC, with the data therein found in *Table 6*.

The highest average values of 5.0 for the year are found both at a latitude facing and at a latitude −15° facing. But further examination of the data shows that the winter values are better at a latitude facing, and this home has an all-electric heat pump system for heating. The slightly better winter performance is attractive to this particular owner. Select the 5.0 average peak sun hours per day value for the calculations, faced at latitude with a fixed tilt.

The equation to determine the actual array capacity is as follows:

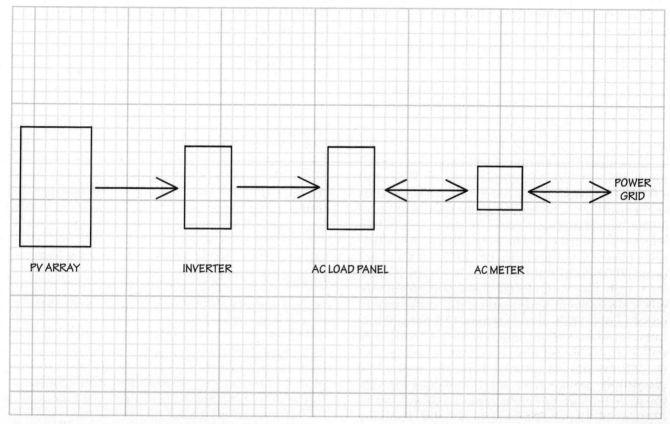

Figure 30 Simple grid-tied system component diagram.

Table 6 Solar Radiation Data for Fixed Tilt Flat Plate Collectors Facing South—Greenville, SC

Tilt Angle		Jan	Feb	Mar	Apr	May	Jun	Jul	Aug	Sep	Oct	Nov	Dec	Avg
	Average	2.6	3.3	4.4	5.6	6.0	6.3	6.0	5.5	4.7	3.9	2.8	2.3	4.3
0°	Minimum	2.2	2.8	3.7	4.6	5.5	5.6	5.3	4.8	4.1	3.3	2.1	1.9	4.2
	Maximum	3.0	4.0	5.1	6.6	6.7	7.2	6.8	6.3	5.3	4.6	3.4	2.6	4.7
	Average	3.5	4.2	5.1	5.9	6.0	6.1	5.9	5.7	5.2	4.8	3.8	3.2	5.0
Lat −15°	Minimum	2.8	3.4	4.1	4.8	5.5	5.4	5.2	4.9	4.4	3.9	2.5	2.5	4.6
	Maximum	4.3	5.3	6.0	7.0	6.7	7.0	6.8	6.6	5.9	5.8	4.8	3.9	5.2
	Average	4.0	4.6	5.3	5.8	5.6	5.6	5.5	5.5	5.2	5.2	4.3	3.7	5.0
Latitude	Minimum	3.1	3.6	4.1	4.7	5.1	5.0	4.8	4.7	4.4	4.2	2.7	2.8	4.7
	Maximum	5.0	5.9	6.3	7.0	6.3	6.4	6.3	6.3	6.1	6.3	5.4	4.6	5.3
	Average	4.2	4.8	5.2	5.4	5.0	4.9	4.8	5.1	5.0	5.2	4.5	4.0	4.8
Lat +15°	Minimum	3.3	3.7	4.0	4.4	4.6	4.3	4.3	4.3	4.2	4.1	2.8	3.0	4.5
	Maximum	5.4	6.1	6.2	6.5	5.6	5.5	5.5	5.8	5.8	6.4	5.8	5.0	5.2
	Average	3.8	3.9	3.6	3.1	2.4	2.2	2.2	2.7	3.2	4.0	3.9	3.6	3.2
90°	Minimum	2.8	3.0	2.8	2.5	2.2	2.0	2.1	2.4	2.7	3.1	2.3	2.7	2.9
	Maximum	4.9	5.1	4.4	3.6	2.6	2.3	2.5	3.0	3.8	5.0	5.1	4.6	3.5

Needed PV system kWh/day ÷ average peak sun hours per day ÷ STC temperature difference correction factor ÷ DC-to-AC combined correction factor = array kW capacity needed

Note that inverter efficiency as an individual factor is not shown in this equation, as was the case in the stand-alone application. This will be discussed further along with other correction factors.

The kWh per day desired has already been determined as 27kWh, with a value of 5.0 average peak sun hours per day. The equation now becomes:

$$\text{Array capacity} = 27\text{kWh/day} \div 5.0 \text{ peak}$$
$$\text{sun hours/day} \div \text{STC temperature difference}$$
$$\text{correction factor} \div \text{DC-to-AC combined}$$
$$\text{correction factor}$$

Remember that standard test conditions (STC) for panels include a temperature of 77°F and 1,000W/m² irradiance. Rarely do these values occur together, which potentially results in unrealistic expectations of panel performance when compared to real operating conditions. An actual cell operating temperature of 122°F (50°C) is not unusual, even in mild weather. For typical panels which lose roughly 0.5 percent performance per °C increase in temperature, this difference in panel temperature accounts for a 12.5 percent loss of performance. Consequently this 12.5 percent value is considered a relatively standard correction factor for most systems. Typically, individual geographic areas have correction values for temperature that are widely accepted. For this example, use the 12 percent correction factor, applying it in the calculations by dividing by 0.088.

The next correction factor in the equation is one which considers a variety of individual factors, combining them mathematically into a single factor called the DC-to-AC combined correction factor. This information and the list of factors considered in *Table 7* is from the NREL and represents factors applied in calculations by the PVWATTS software program used by many designers. Note that each factor has a range of potential values, with the default value used in the program listed in the center column.

The designer may need to make some adjustments in these factors to support a specific situation. The following is a brief description of each factor:

- *PV panel nameplate DC rating* – This value accounts for the inaccuracy of the manufacturer's nameplate rating. A correction factor of 0.95 indicates that independent testing yielded power measurements at STC that were 5 percent less than the manufacturer's nameplate rating.
- *Inverter and transformer* – This indicates the combined loss of efficiency at both the inverter and transformer. Note that inverter efficiency was considered separately in the stand-alone system calculations. The designer generally works with the same line of inverters regularly and is familiar with the ratings, even though the final selection may not have been made at this point.

Table 7 DC-to-AC Combined Correction Factor Components

Component Correction Factors	PV Watts Default	Range
PV panel nameplate DC rating	0.95	0.80–1.05
Inverter and transformer	0.92	0.88–0.98
Mismatch	0.98	0.97–0.995
Diodes and connections	0.995	0.99–0.997
DC wiring	0.98	0.97–0.99
AC wiring	0.99	0.98–0.993
Soiling	0.95	0.30–0.995
System availability	0.98	0.00–0.995
Shading	1.00	0.00–1.00
Sun-tracking	1.00	0.95–1.00
Age	1.00	0.70–1.00

- *Mismatch* – The correction factor for PV panel mismatch accounts for manufacturing tolerances that result in PV panels with slightly different current-voltage characteristics. As a result, they do not operate at their peak efficiencies. The default value of 0.98 represents a loss of 2 percent because of mismatch.
- *Diodes and connections* – This factor accounts for losses from voltage drops across diodes used to block the reverse flow of current and from resistive losses in electrical connections.
- *DC wiring and AC wiring* – These two factors account for simple resistive losses in the AC and DC interconnecting wiring.
- *Soiling* – Soiling from dirt, snow, and other foreign matter on the surface of the PV panel prevents solar radiation from reaching the cells. Soil accumulation is generally location and weather dependent. The spring of 2010 was one of the worst in decades for levels of airborne pollen across the country—one significant source of panel fouling. There are typically greater soiling losses in high-traffic, high-pollution areas with infrequent rain. For northern locations, snow often reduces the energy produced. Low PV array tilt angles prevent snow from sliding off and impedes the washing effect from rains, contributing to the problem. Testing revealed that a roof-mounted PV system in Minnesota with a tilt angle of 23 degrees experienced a 70 percent reduction in winter power

NCCER — *Contren® Learning Series* 57103-11

production due to consistent snow, while a nearby array tilted at 40 degrees experienced a 40 percent reduction.

- *System availability* – Correcting for system availability accounts for times when the system is off because of maintenance, repair, or utility outages. The default value of 0.98 represents the system being off two percent of the year.
- *Shading* – Tools such as the Solar Pathfinder™ can determine a correction factor for shading by buildings and objects. Obviously, shading must be avoided, as previously discussed.
- *Sun-tracking* – The factor for sun-tracking accounts for losses from one- or two-axis tracking systems when the tracking mechanisms are not 100 percent effective in tracking the sun. In the selection of the average peak sun hours for the example system, choose the value for fixed arrays that already accounts for certain losses. Unless there is a specific reason to believe there will be additional losses due to poor or nonexistent sun tracking, this value should remain at 1.00.
- *Age* – The loss in panel performance is typically one percent per year due to age alone. To calculate the array performance based on future years, the designer can account for age by deducting 0.01 per year of age.

The list of small factors affecting the array can become one highly significant factor in final performance. The final factor is calculated by multiplying all of the values together:

$$0.95 \times 0.92 \times 0.98 \times 0.995 \times 0.98 \times 0.99 \times$$
$$0.95 \times 0.98 \times 1.00 \times 1.00 \times 1.00 = 0.77$$

If changes are made to any of the default values shown, simply recalculate for the new final value. The example calculations use the PV watts default value.

Returning to the equation for sizing the array, insert the correction factors and determine the final size:

Needed PV system kWh/day ÷ average peak sun hours per day ÷ STC temperature difference correction factor ÷ DC-to-AC combined correction factor = PV kW capacity needed

27kWh/day ÷ 5.0 average peak sun hours/day ÷ 0.88 loss due to temperature ÷ 0.77 DC-to-AC combined correction factor = 7.97kW capacity needed from the array. This can also be expressed as 7,970W.

Figure 31 provides technical data on one model of solar panel produced by Canadian Solar—the CS5P monocrystalline panel. If selecting a 220W panel, 36 panels will provide 7,920W. This would be a good choice for this project. Note that these panels are nominal 48 VDC models.

Figure 32 provides a summary of the panel and array selection.

4.1.2 Inverter Selection

Inverters designed for use in grid-connected systems are required by the *NEC*® to meet two specific standards of construction and testing: *UL Standard 1741* and *IEEE Standard 1547*. These national standards have resulted in inverters

ELECTRICAL DATA		CS5P-220M	CS5P-225M	CS5P-230M	CS5P-235M	CS5P-240M	CS5P-245M	CS5P-250M
Nominal Maximum Power at STC (Pmax)		220W	225W	230W	235W	240W	245W	250W
Optimum Operating Voltage (Vmp)		47.0V	47.4V	47.5V	47.7V	48.1V	48.4V	48.7V
Optimum Operating Current (Imp)		4.68A	4.74A	4.84A	4.93A	4.99A	5.06A	5.14A
Open Circuit Voltage (V_{OC})		58.8V	59.0V	59.1V	59.2V	59.4V	59.5V	59.6V
Short Circuit Current (I_{sc})		5.01A	5.09A	5.18A	5.27A	5.34A	5.43A	5.49A
Operating Temperature		−40°C~+85°C						
Maximum System Voltage		1,000V (IEC)/600V (UL)						
Maximum Series Fuse Rating		10A						
Power Tolerance		+5W						
Temperature Coefficient	Pmax	−0.45%/°C						
	V_{OC}	−0.35%/°C						
	I_{sc}	−0.60%/°C						
	NOCT	45°C						

Under Standard Test Conditions (STC) of irradiance of 1000 W/m³, spectrum AM 1.5, and cell temperature of 25°C

103F31EPS

Figure 31 Canadian Solar CS5P electrical data.

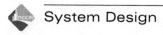

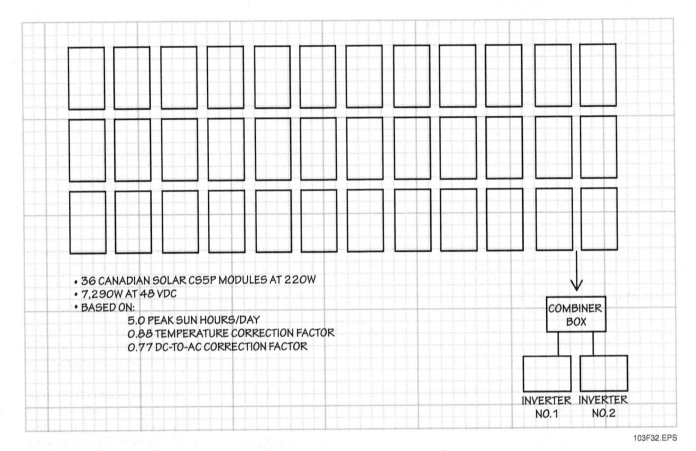

- 36 CANADIAN SOLAR CS5P MODULES AT 220W
- 7,290W AT 48 VDC
- BASED ON:
  - 5.0 PEAK SUN HOURS/DAY
  - 0.88 TEMPERATURE CORRECTION FACTOR
  - 0.77 DC-TO-AC CORRECTION FACTOR

COMBINER BOX

INVERTER NO.1   INVERTER NO.2

103F32.EPS

Figure 32 Example grid-tied array selection.

which perform consistently and with a high level of safety and protection.

When sizing and selecting the inverter for the stand-alone application, the calculations focused on the total connected AC load. However, the inverter sizing for a grid-tied application focuses on the maximum amount of power that will be routed through the inverter from the array, since the system will also be sending PV power to the grid. Depending upon the design, the system may be capable of sending power to the grid even when the local AC load is at its highest. The controlling factor for the work of the inverter then becomes the power the array generates at its maximum capacity, not just the local AC load alone.

The calculation to determine the total current from the array will be based on the wattage output at STC. Remember that these panels have already been capacity-corrected in earlier calculations for all potential variations from laboratory and STC conditions. The 220W panels then, with 36 panels at work, should produce 7,920W at STC. The inverter must have a continuous power rating of at least 7,920W.

This exercise uses data from the Fronius line of grid-connected inverters to determine which product would be appropriate for the application. Begin the selection process by examining the nominal rating of the units, looking for a size that meets the capacity requirements. Refer to *Table 8*.

Table 8 Fronius IG Series Basic Specifications

Inverter	Max AC Output	Nom AC Output	AC Voltage	DC Input Voltage
IG 2000	2,000W	1,800W	240V	150–500V
IG 3000	2,700W	2,500W	240V	150–500V
IG 2500-LV	2,350W	2,150W	208V	150–500V
IG 4000	4,000W	4,000W	240V	150–500V
IG 5100	5,100W	5,100W	240V	150–500V
IG 4500-LV	4,500W	4,500W	208V	150–500V

The Fronius IG line offers inverters nominally rated up to 5,100W AC maximum output. In the IG Plus line, larger inverters are available, but nothing that quite fits the application of ~8kW, since the line skips from 7.5kW to 10kW. The 10kW model could be used, but remember that efficiency will likely be reduced somewhat as a result.

This application offers a good opportunity to parallel two inverters. Again, examine the models shown in *Table 8*. Using two IG4000 models would match the needed capacity of 7,920W nicely. This unit cannot yet be considered the final selection though, based solely on nominal capacity. Additional factors must be examined to ensure it is compatible with the overall design and strategy.

Grid-connected inverters such as these have quite a few specifications to consider. One very important requirement which significantly impacts the project is the inverter's DC voltage operating range. Note that the required DC voltage for all the inverters in the Fronius IG model line is 150V to 450V, with a maximum input value of 500 VDC. The challenge then, is to ensure the array, as presently designed, can be arranged in appropriate series configurations of individual panels to consistently provide the required voltage to the inverter. If the inverter begins to receive less than 150 VDC from the connected panels, its operation ceases.

MPPT capabilities are also available in grid-tied inverters. You will recall that charge controllers incorporated this capability to take advantage of as much array power as possible in the battery charging process. For grid-tied inverters, the goal is to regulate the array voltage at the maximum power point value shown on the panel I-V curve. This feature has become quite common on inverters.

The best method to determine the proper number and size of the panel strings clearly is to consult the manufacturer's literature. Many, including Fronius, offer on-line software to assist in selection and design issues such as this. After accessing the Fronius Configuration Tool and downloading the software, enter the chosen inverter model. *Figure 33* provides a screenshot of the example system parameters inserted into the program. This particular tool is really quite simple, as it already lists all common panel manufacturers and the individual models in the product line. After selecting the desired panels (in this case, Canadian Solar CS5P-M), the program provides possible panel configurations.

Remember that the component selection produced two equally sized inverters. Each inverter therefore, must handle 3,960W from the array, representing 18 of the 36 PV panels. By consulting the configuration table in *Figure 34*, determine if there is a possible configuration using 18 panels that will supply the correct voltage consistently. Yellow boxes indicate that the inverter may be oversized, while green boxes represent ideal configurations. Note that one of the green options represents a configuration of three series strings of six panels each, a perfect combination for the 18 available panels which also accommodates the 3,960 watt production from one half of the total array. Choosing this configuration with a mouse click results in additional electrical information being revealed to the designer. These results are then easily printed using the Print button at the bottom of the page. The printed report is shown in *Figure 34*.

It is important to note that this is only one example of the many possible inverters and design software offerings available to the designer. As discussed previously, the experienced designer will become quite familiar with the products used regularly and comprehend any unique details regarding their selection process and application.

The array and inverter selections are the areas where grid-tied system component selection differs most from stand-alone systems. Other aspects of the design, such as wire sizing and array mounting, are common among the two approaches. Now that the primary components have been selected, it is necessary to review some of the details associated with the grid interface process.

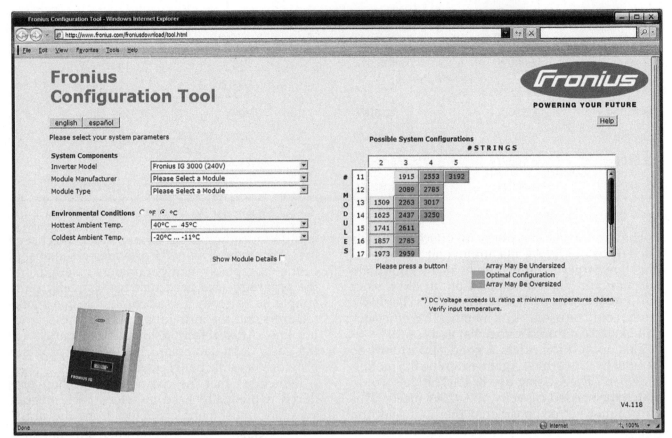

Figure 33 Fronius configuration tool screenshot.

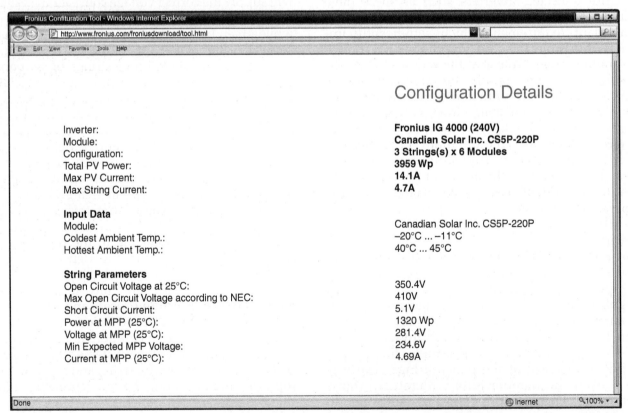

Configuration Details

Inverter:	**Fronius IG 4000 (240V)**
Module:	**Canadian Solar Inc. CS5P-220P**
Configuration:	**3 Strings(s) x 6 Modules**
Total PV Power:	**3959 Wp**
Max PV Current:	**14.1A**
Max String Current:	**4.7A**

Input Data

Module:	Canadian Solar Inc. CS5P-220P
Coldest Ambient Temp.:	−20°C ... −11°C
Hottest Ambient Temp.:	40°C ... 45°C

String Parameters

Open Circuit Voltage at 25°C:	350.4V
Max Open Circuit Voltage according to NEC:	410V
Short Circuit Current:	5.1V
Power at MPP (25°C):	1320 Wp
Voltage at MPP (25°C):	281.4V
Min Expected MPP Voltage:	234.6V
Current at MPP (25°C):	4.69A

Figure 34 Fronius configuration tool inverter selection report.

4.2.0 PV System Grid Interface

Grid-connected PV systems can be connected to either side of the owner's utility service disconnect. PV systems owned by the user (as opposed to utility ownership) are typically connected on the load side of the meter. The details of the connection and the utilities' requirements are part of the interconnection agreement.

The utility providing service to the example location in South Carolina is Duke Power. The Public Service Commission of SC approved the *Standard for Interconnecting Small Generation 100KW or Less with Electric Power Systems* effective at the end of 2006, in an effort to streamline the interconnection process. A short checklist outlines the necessary steps, including the need to submit a simple, one-line diagram of the system to be installed, a detailed list of the equipment selected, and the desired metering arrangement (net metering versus a direct grid connection and separate meter to quantify power generated and sold). A generating system which seeks to sell electricity to Duke Energy Carolinas under a Purchased Power Agreement must be a Qualifying Facility as defined by the *Public Utility Regulatory Policies Act of 1978 (PURPA)* and the Federal Energy Regulatory Commission (FERC) regulations implementing PURPA. A copy of the FERC Qualifying Facility Certificate filing should accompany the interconnection application.

The above is provided as an example of the process for one state and one utility only. The requirements and documents vary from state-to-state, and utility-to-utility, but the same issues are commonly encountered. A system designer may advise the building owner of the requirements and documentation involved. However, it is important to remember that interconnection agreements are between the owner/consumer and the utility, not the contractor or installer. At a minimum, detailed information from the design must be documented on the application.

4.2.1 Net Metering and the Local Utility

Net metering represents the simplest and most advantageous method of monitoring power production and use in a grid-tied system. As a grid-tied array produces power in excess of that being used on site, that power is fed back to the grid through the meter. With power passing in the opposite direction, the meter registers it as a credit. In the end, if the user consumes 3kW of power overnight, but produces 3kW of excess power fed back to the grid during the day, then the meter registers net zero for the period. This simple method eliminates any need for trying to separately meter PV-produced power and utility usage, then attempting to reconcile the values mathematically. Net metering essentially allows the PV system to store free power in the grid for later use, with the potential to actually provide some utility revenue to the operator in lieu of a utility bill. The final financial outcome depends entirely on the system capacity and the power needs of the application.

Discussing the system installation and grid interface process with the local utility in the beginning design stage is vital. Before the grid interface is created, an interconnection agreement with the utility is required. The agreement will specify the terms and conditions of power buying and selling, and outline the technical requirements of the connection itself. For example, most interconnection agreements require that the PV system be insured by the owner to protect the utility from damage caused by the PV system. Although national standards are in development, they are not mandated and, therefore, individual utilities may differ in their interconnection agreement requirements. After a period of designing and installing PV systems which interface with a particular utility, the designer will become familiar with not only their standards and expectations, but the personnel involved as well. This will facilitate the interface process and allow a smooth transition into the new relationship between consumer and utility.

One nonprofit organization in the U.S. has developed a variety of standards and provides a national forum for interested parties to exchange information and develop best practices for the power industry. The Interstate Renewable Energy Council (IREC) mission includes "creating renewable energy programs and policies targeted at the adoption of uniform guidelines, standards and quality assessment." The IREC's *Model Interconnection Standard and Net Metering Rule* (updated in 2009) is considered to be the primary resource available on current best practices. These model procedures serve as a template for regulators and utilities to develop or update local and state standards.

4.2.2 NEC® Requirements for the Point of Connection

Typically, the inverter AC output will be connected to what would normally be considered the load side of a dedicated circuit breaker in the main distribution panel. The need for a dedicated circuit breaker, or a fusible disconnect, at the point of connection is determined from *NEC Section 690.64(B)*. The circuit breaker used for this application must be suitable for the purpose. Per *NEC Section 690.54*, the point of interconnection with the grid must be clearly marked in the vicinity of the disconnecting means as being a source of power, and indicate the rated AC output current and nominal AC voltage.

The sum of the overcurrent protective device ratings which feed power to the panelboard must not exceed 120 percent of the busbar rating in all residential or commercial PV system installations. For example, a 200A panel, fed from the grid through a 200A main breaker, can accommodate up to a 40A circuit breaker from an inverter power source. The 240A total overcurrent protective device rating of the inputs equals 120 percent of the panelboard rating. A relatively small PV system can therefore be connected without the panelboard being replaced with a larger size. However, the designer and installer must remember that the overcurrent protective device for the inverter input will be sized at 125 percent of the inverter's rated output current. As a result, if a 40A circuit breaker for the inverter input is the largest allowed, an inverter with an output rating of 32A would be the largest inverter that could be connected to the busbar.

Taking this issue further, remember that a 120 VAC inverter would be connected to a single busbar in a panelboard. In this case, the rule applies exactly as explained above. But a 240 VAC connection involves connection to two busbars. That means the output rating of the inverter can be twice as large. Two inverters, can also be used in this situation, with each one connected to each busbar through a two-pole circuit breaker.

Unless certain exceptions are met per *NEC Section 705.12*, the point of connection must be on the line side of all ground fault protection devices.

There is no replacement for intensive study and research into local practices, utility policies, and code requirements for the specific location of the PV system. Once a designer has experienced several PV system installations in a given locale, the design process and meeting the electrical requirements for safe and reliable operation of the system will seem much less complex.

SUMMARY

The design phase of a solar PV project is crucial to the success of the finished product. It requires a great deal of thought, experience, and a thorough understanding of the individual components to be used. The designer must rely on a complete and thorough site assessment to develop the best approach. Further, even the best designs can easily be rendered unsuccessful by a poor installation. In essence, all phases of the solar PV project are interdependent, and a poor result at any level will impact the entire project negatively.

A designer must first evaluate the big picture and conceive an overall design based both on site information and the goals of the owner. Once the overall approach has been determined, major components are selected and their locations determined on the site. After these basic elements are in place, the majority of the remaining work involves tasks such as the selection of appropriate interconnecting wire sizes and types, overcurrent protective devices, and power disconnect components. Once all aspects of the design phase have been considered and the basic layout complete, a material takeoff can begin and job costing processes initiated.

1. The types of electrical loads that PV systems can provide power for include _____.
   a. only DC electrical loads
   b. only AC electrical loads
   c. only those loads which operate during the day
   d. both AC and DC loads

2. Using the equation Qty × volts × amps = AC watts × hrs/day × days/week ÷ 7 days/week = AC Wh per day, determine the daily AC Wh load to be considered for one 120VAC device which draws 4.5A and is in operation 9 hours per day, 5 days per week.
   a. 540Wh
   b. 3,471Wh
   c. 4,860Wh
   d. 24,300Wh

3. Changing the DC system design voltage from 12V to 48V has the effect of _____.
   a. increasing current flow by a factor of 2
   b. increasing current flow by a factor of 4
   c. decreasing current flow by 50 percent
   d. decreasing current flow by 75 percent

4. If a solar system battery is rated at a nominal capacity of 220Ah, how many hours can it be expected to deliver power at a consistent load of 11A with a 50 percent depth of discharge?
   a. 2 hours
   b. 5 hours
   c. 10 hours
   d. 20 hours

5. Referring to the I-V curve, solar panels produce their maximum power at what point of the curve?
   a. The highest value on the current axis
   b. The knee of the curve
   c. The highest value on the voltage axis
   d. The point where the DC system design voltage crosses the curve

6. Single-axis tracking systems for arrays track _____.
   a. by reading solar insolation dynamically
   b. the sun's altitude
   c. the sun's azimuth
   d. solar position by panel temperature

7. During the bulk charge stage, batteries have typically reached a charge level of _____.
   a. 45 to 55 percent
   b. 75 percent
   c. 80 to 90 percent
   d. 95 percent

8. The identifying feature of a diversion controller is that _____.
   a. it pulses charging power to the batteries
   b. it opens the battery charging circuit once the process is complete
   c. array power is used to power another DC load once charging is complete
   d. each battery is charged separately, one at a time

9. The information needed to select a non-MPPT controller includes the _____.
   a. number of batteries in the bank
   b. number of solar panels wired in parallel
   c. maximum AC current load
   d. desired amount of time a full charge will require

10. Advances in inverter design and construction have allowed them to reach efficiency levels as high as _____.
    a. 99 percent
    b. 95 percent
    c. 92 percent
    d. 90 percent

11. The primary information required for selection of an inverter for a stand-alone PV system is _____.
    a. AC load wattage, individual solar panel voltage, AC output voltage, and AC frequency
    b. AC load wattage, DC input voltage, AC output voltage, and the array short circuit current
    c. DC load wattage, DC output voltage, AC input voltage, and AC frequency
    d. AC load wattage, DC input voltage, AC output voltage, and AC frequency

12. When determining the wire sizing ampacity for the connection of power from the solar combiner box to either a controller or inverter, a unique multiplier of 1.56 is applied to the array short circuit current to _____.

a. compensate for an anticipated long wiring run
b. allow for future expansion of the array
c. compensate for the exposure to sunlight
d. derate to 80 percent plus compensate for periods of extreme insolation

13. The device used to collect battery bank power into a single, noninsulated conductor is _____.

a. a busbar
b. the charge controller
c. the service cable
d. the combiner

14. The wire size for the inverter AC output is based on the _____.

a. maximum power flow from the array
b. maximum power flow from the battery bank, when present
c. inverter's maximum rated power output at standard test conditions
d. inverter's maximum output at 40°C

15. Using the inverter's lowest acceptable DC voltage for sizing the DC input wiring _____.

a. prevents the wire size from being larger than necessary
b. ensures the wire size is large enough to handle the current if the DC input voltage falls lower than design
c. is not recommended by the *NEC*®
d. increases the efficiency of the inverter

16. The *NEC*® specifies that arrays which are designed to provide voltages in excess of 50V _____.

a. can be protected by a single overcurrent protective device
b. be installed with a means of disconnect for each individual panel
c. be labeled as a high-voltage array
d. incorporate panel fuses as well as an overcurrent protective device for the full array

17. In stand-alone PV systems, the main breaker of the load panel _____.

a. will be sized based on the same ampacity value used to size the inverter AC output wiring
b. is sized to handle the total of the connected loads
c. is sized to match the capacity of the panel
d. will be sized to the total of the connected loads, multiplied by 1.25

18. Per the *NEC*®, solar panels must be connected in a manner which _____.

a. allows removal of an individual panel while the system is in operation
b. allows for the removal of an individual panel without interruption of a grounded conductor from another circuit
c. ensures the first and last panels produced balanced power
d. prevents the output voltage from dropping below 12V

19. Where grounding of a conductor is planned for a system exceeding 50VDC, _____.

a. the positive conductor is typically chosen
b. it can be grounded at several strategic locations
c. it must be grounded at only one point
d. the conductor color must be green

20. Equipment grounding is required _____.

a. on all systems operating in excess of 50VDC
b. on all systems operating in excess of 15VDC
c. on grid-tied systems only
d. on all systems, regardless of voltage

21. The *NEC*® requires that when an inverter is removed from the circuit for service or other reason, _____.

a. another inverter must be immediately installed in its position
b. jumper wires must be present to preserve the continuity of equipment and system ground wiring
c. the entire PV system must be disabled
d. all DC input wiring must be physically disconnected from its source

22. Solar panels degrade in power production as their temperature increases above STC of 25°C, typically at a rate of _____.
    a. 0.5 percent per °C
    b. 1 percent per °C
    c. 2 percent per °C
    d. 2.5 percent per °C

23. When calculating the needed solar array capacity for a grid-tied system without batteries, the efficiency of the inverter is _____.
    a. not a factor that needs to be considered
    b. a specific factor in the equation
    c. compensated for as part of the DC-to-AC correction factor
    d. part of the mathematical calculation for the inverter selection process

24. One provision which is likely to be found in an interconnection agreement is _____.
    a. the specified solar panel brand and size that is acceptable to the utility
    b. that the owner must maintain insurance against potential damage or injury the system could cause
    c. the maximum inverter capacity allowed
    d. the frequency of system inspections to be made by the utility

25. An inverter's rated output current is 60A. What is the maximum size overcurrent protective device that can be installed?
    a. 48A
    b. 52A
    c. 60A
    d. 75A

Trade Terms Quiz

1. An agreement that specifies the terms and conditions of power buying and selling between the utility and the PV system owner or operator and outlines the technical requirements of the PV system connection interface is called the _____.

2. A battery charge controller that opens the charging circuit from the PV array once charging is complete is a(n) _____.

3. A special charging cycle used intermittently on flooded batteries to rejuvenate and circulate the electrolyte through gassing by applying higher than normal charge voltages for a specified period of time to create a bubbling activity inside the battery and redistribute the electrolyte is called the _____.

4. An application-specific junction box in which multiple solar panels are connected together to combine their output before being routed to the charge controller or inverter is the _____.

5. When one conductor of a two-conductor system is connected to ground, it is referred to as _____.

6. The recommended lowest level of battery discharge, measured in volts and specified by the manufacturer for a given battery type and construction, which is used as a control setpoint to disconnect the battery from the load to prevent its permanent damage, is called the _____.

7. The second stage of battery charging, where the voltage remains constant, current is gradually reduced as resistance in the circuit increases, and the charging voltage is typically highest, is the _____.

8. The style of battery charge controller that diverts PV array capacity to another DC load whenever the battery charging process is inactive is called a(n) _____.

9. When an imbalance is detected between the current passing through a hot conductor and that passing through the neutral conductor, indicating current flow to ground, the _____ actuates to open the circuit.

10. A simple form of battery charge controller that shorts the PV array by charging power to a false load when the charging process is complete, but creates heat in the process, is the _____.

11. The number of days that a solar PV system can continue to meet an electrical load requirement without additional input from the solar array is called its _____.

12. The voltage setpoint at which a battery charge controller perceives roughly 80 to 90 percent of the battery's fully charged state and initiates the absorption stage is the _____.

13. The chemical process inside a battery that causes large crystals of lead sulfate to form on the plates instead of desirable minute crystals is called _____.

14. To generate the desired voltage and current capacity, a collection of batteries is electrically connected in parallel and series combinations to form a(n) _____.

15. When a conductor connects all non-interruptible, noncurrent-carrying components of an installation to the chassis or frame of an appliance or unitary piece, it is called _____.

16. Often referred to as being at trickle charge or maintenance charge, the third stage of three-stage battery charging, where charging voltage is reduced to maintain the fully charged condition of the battery is the _____.

17. The initial stage of three-stage battery charging, where the maximum amount of current is delivered to the battery until it reaches 80 to 90 percent of its charge capacity is the _____.

Trade Terms

Absorption stage	Diversion controller	Interconnection agreement	Single-stage controller
Battery bank	Equalization stage		Solar combiner
Bulk charge stage	Equipment grounding	Low-voltage disconnect (LVD)	Sulfation
Bulk voltage setting	Float stage		System grounding
Days of autonomy	Ground fault protection	Shunt controller	

L. J. LeBlanc

Senior Electrician, Photovoltaic Instructor
Pumba Electric LLC
Port Allen, LA

Listen to your peers. Learn from others' experience. What's most important is what you learn after you think you know it all.

How did you get started in the construction industry?
My dad got me into this field. He was educated in electrical work at the Coyne American Institute in Chicago, and he had me wiring houses from the time I was nine years old. First I worked as a helper and then I began an apprenticeship through trade school. There was a lot of work then. Within a year I was made foreman. At one point, I took a two-week job that lasted fourteen years.

What do you enjoy most about your work?
Electrical work is challenging. There's always something new to work out, a new problem to solve. If a man made it, I can fix it.

Do you think training and education are important in construction? Why?
Yes, you need to get a good education and finish it with on-the-job training. Listen to your peers. Learn from others' experience. What's most important is what you learn after you think you know it all.

How important do you think NCCER credentials are to workers today?
Anything you can take with you from one location to another is good. It shows you've had the training.

How have training and construction impacted your life and career?
I have great respect for good training combined with experience. Today I do all kinds of electrical work, estimating, bidding, installation—even troubleshooting. I like walking in to something different every day. I'm still learning and don't plan on stopping.

Would you suggest construction as a career to others?
Well, I encouraged my son to go into it, and now I work for him, so yes, I recommend it. If you work hard and keep up with the technology, there'll always be a demand for your services.

Do you have any advice for people just entering the field?
I believe in giving eight hours' work for eight hours' pay, and I'm never late. If I can't be there on time, I'm there early. That's worked well for me, and I think it would work for the next generation too.

How do you define craftsmanship?
You sign your name with your work. It not only represents you but your company.

Trade Terms Introduced in This Module

Absorption stage: The second stage of battery charging, where the voltage remains constant and current is gradually reduced as resistance in the circuit increases. This stage continues until a full charge condition is sensed. During this stage, the charging voltage is typically highest, from roughly 14V to 15.5V.

Battery bank: A collection of batteries electrically connected in parallel and series combinations to generate the desired voltage and current capacity needed.

Bulk charge stage: The initial stage of three-stage battery charging, where the maximum amount of current is delivered to the battery until it has reached 80 to 90 percent of its possible charge capacity. Voltage generally ranges from 10.5V to 15V through this stage.

Bulk voltage setting: The voltage setpoint for a battery charge controller to perceive as roughly 80 to 90 percent of the battery's fully charged state, initiating the absorption stage.

Days of autonomy: The number of days that a solar PV system can continue to meet an electrical load requirement without additional input from the solar array.

Diversion controller: A style of battery charge controller which diverts PV array capacity to another DC load whenever the battery charging process is inactive.

Equalization stage: A special charging cycle used intermittently on flooded batteries to rejuvenate and circulate the electrolyte through gassing. Higher than normal charge voltages are applied for a specified period of time, creating a bubbling activity inside to redistribute the electrolyte.

Equipment grounding: Connecting the chassis or frame of an appliance or unitary piece to ground. This ground should connect all components of an installation which do not conduct current and should be uninterrupted.

Float stage: The third stage of three-stage battery charging, where charging voltage is reduced to maintain the fully charged condition of the battery. This is often referred to as at trickle charge or maintenance charge.

Ground fault protection: The incorporation of devices which monitor the passage of current in a hot conductor and compares it to that passing through the neutral conductor. An imbalance indicates current flow to ground and the ground fault protection device actuates to open the circuit.

Interconnection agreement: An agreement that specifies the terms and conditions of power buying and selling between the utility and the PV system owner or operator, and outlines the technical requirements of the PV system connection interface.

Low-voltage disconnect (LVD): The recommended lowest level of battery discharge, measured in volts, and specified by the manufacturer for a given battery type and construction. This value is used as a control setpoint to disconnect the battery from the load to prevent its permanent damage.

Shunt controller: A simple form of battery charge controller which shorts the PV array charging power to a false load when the charging process is complete, creating heat.

Single-stage controller: A battery charge controller which opens the charging circuit from the PV array once charging is complete. Most incorporate other features, such as float stages, once this level of controller design is reached.

Solar combiner: An application-specific junction box where multiple solar panels are connected together, combining their output together before being routed to the charge controller or inverter.

Sulfation: A chemical process inside a battery which causes large crystals of lead sulfate to form on the plates, instead of desirable minute crystals. This process generally occurs in batteries which experience poor charging cycles or remain discharged.

System grounding: Connecting one conductor of a two-conductor system to ground.

Appendix A

STAND-ALONE SYSTEM COMPONENT SELECTION FORMS

STAND-ALONE SYSTEM – LOAD DETERMINATION

AC Load	Qty	× Volts	× Amps	= AC Watts	× Hrs/Day	× Days/Week	÷ 7 Days/Week	= AC Watt/Hours

Total AC Watts = Total AC Avg W/H Load Daily =

DC Load	Qty	× Volts	× Amps	= DC Watts	× Hrs/Day	× Days/Week	÷ 7 Days/Week	= DC Watt/Hours

Total DC Watts = | **Total DC Avg W/H Load Daily =**

BATTERY BANK SIZING

AC average daily load = _____ watt-hours per day

Expected inverter efficiency (decimal equivalent) = _____

DC average daily load = _____ watt-hours per day

DC system design voltage = _____ volts DC

Days of autonomy = _____ days

Desired battery discharge limit (decimal equivalent) = _____

Chosen battery = _____ AH capacity = _____ amp/hours

Chosen battery voltage = _____ volts DC

AC Average Daily Load	×	Inverter Efficiency	=	Corrected AC Average Load	+	DC Average Daily Load	=	W/H Load
_____		_____		_____		_____		_____

W/H Load	÷	DC System Design Voltage	=	Average A/H per Day
_____		_____		_____

Average A/H per Day	÷	Days of Autonomy	÷	Discharge Limit	÷	Battery A/H Capacity	=	Batteries in Parallel
_____		_____		_____		_____		_____

DC System Voltage	÷	Battery Voltage	=	Batteries in Series	×	Batteries in Parallel	=	Total Batteries
_____		_____		_____		_____		_____

ARRAY SIZING

Average amp/hours per day = _____ A/H per day

Chosen battery efficiency (decimal equivalent) = _____

Peak sun/hours per day = _____

Chosen PV module = _____

    – Maximum power current (I_{mp}) = _____ amps

    – Short circuit current (I_{sc}) = _____ amps

    – Nominal module voltage = _____ volts DC

Average A/H per Day	÷	Battery Efficiency	÷	Peak Sun Hours per Day	=	Array Peak Amps
_____		_____		_____		_____

Array Peak Amps	÷	Maximum Power Current (I_{mp})	=	Modules in Parallel
_____		_____		_____

DC System Voltage	÷	Nominal Module Voltage	=	Modules in Series	×	Modules in Parallel	=	Total Modules
_____		_____		_____		_____		_____

System Design

CONTROLLER SELECTION

Chosen PV module short circuit current (I_{sc}) = _____ amps

Modules in parallel (from array sizing form) = _____

Total DC load watts (from load determination form) = _____ watts DC

DC system voltage = _____ volts DC

Important/desired controller features and options:

Module Short Circuit Current (I_{sc})	×	Modules in Parallel	×	1.25 Safety Factor	=	Array Short Circuit Amps Controller Capacity Required
_____		_____		1.25		_____

Total DC Load Watts	÷	DC System Voltage	=	Total DC Load Amps Controller Capacity Required
_____		_____		_____

Selected controller make, model, and specifications:

INVERTER SELECTION

AC total connected wattage = _____ watts capacity required

AC projected surge wattage = _____ watts surge capacity required

AC design voltage output = _____ volts AC output required

Frequency = _____ Hz required

DC system input voltage = _____ volts DC input required

Important/desired inverter features and options:

Selected controller make, model, and specifications:

GRID-TIED SYSTEM COMPONENT SELECTION FORMS

GRID-TIED ARRAY SIZING

Yearly average energy consumption = _____ kWh, ÷ 365 days = _____ kWh/day

Desired kWh per day from solar (decimal equivalent) = _____

_____ kWh/day × _____ desired kWh from solar = _____ PV system kWh/day

---

PV system kWh/day = _____

Average peak sun/hours per day for location = _____

Temperature difference correction factor (TDC) = _____ (see Page 2)

DC-to-AC combined correction factor (DCACCC) = _____ (see Page 2)

Chosen PV module = _____

    – STC Watt rating = _____ watts

PV System kWh/day	÷	Avg Peak Sun Hours/Day	÷	TDC	÷	DCACCC	×	1,000	=	Required PV Array Watts
_____		_____		_____		_____		1,000 1,000		_____

Required PV Array Watts	÷	Module STC Watt Rating	=	Total Modules Required
_____		_____		_____

TEMPERATURE DIFFERENCE CORRECTION FACTOR

Based on a typical performance loss of 0.5% per degree °C in PV modules that results from increased temperature, calculate the TDC by:

- Determining the anticipated module temperature
- Determining the temperature difference between your value and the STC value of 25°C
- Multiply the resulting temperature difference by 0.5% to find the percentage of loss

Example: 40°C projected module temperature −25°C STC value = 15°C × 0.5% = 7.5% loss, converted to a decimal multiplier = 0.925 multiplier

DC-TO-AC COMBINED CORRECTION FACTOR

Component Correction Factors	Default	Chosen Value	Range
PV module nameplate DC rating	0.95		0.80–1.05
Inverter and transformer	0.92		0.88–0.98
Mismatch	0.98		0.97–0.995
Diodes and connections	0.995		0.99–0.997
DC wiring	0.98		0.97–0.99
AC wiring	0.99		0.98–0.993
Soiling	0.95		0.30–0.995
System availability	0.98		0.00–0.995
Shading	1.00		0.00–1.00
Sun-tracking	1.00		0.95–1.00
Age	1.00		0.70–1.00

Product of chosen correction factors = 0.96001–0.09999

Choose a correction factor value for each line item, then multiply to determine the DC-to-AC combined correction factor.

Using the default values as an example, the factor would be:

$$0.95 \times 0.92 \times 0.98 \times 0.995 \times 0.98 \times 0.99 \times 0.95 \times 0.98 \times 1.00 \times 1.00 \times 1.00 = 0.77 \text{ DCACCC}$$

GRID-TIED INVERTER SELECTION

Number of PV modules determined to be = _____

Chosen module wattage @ STC = _____ watts

Number of PV Modules	×	STC Wattage	=	Maximum Wattage Inverter Throughput Required
_____		_____		_____

Maximum Wattage Inverter Throughput Required	÷	Number of Desired Inverters	=	Maximum Wattage Needed per Inverter
_____		_____		_____

Number of PV Modules	÷	Number of Inverters	=	Number of Modules per Inverter
_____		_____		_____

Preferred inverter(s) make, model, and basic specifications:

Determine the number of modules required in series for the chosen inverter to operate comfortably within the DC input voltage range. This is done through review of the manufacturer's literature, or through the use of manufacturer-provided selection software. Ensure that the configuration of series-parallel modules in the array, or subarrays, matches the inverter specifications appropriately. When mismatches exist, adjustments must be made in the inverter model chosen, the PV module wattage chosen, or the number of modules in the array.

For selection software use, the highest and lowest expected ambient temperatures for the region as well as the module specifications will be needed.

Additional Resources

This module presents thorough resources for task training. The following resource material is suggested for further study.

IEEE 1547, Standard for Interconnecting Distributed Resources with Electric Power Systems, Latest Edition. Los Alamitos, CA: Institute of Electrical and Electronics Engineers (IEEE).

National Electrical Code® (NFPA 70®), Latest Edition. National Fire Protection Association (NFPA): Quincy, MA.

Occupational Safety and Health Standard 1910.302, Electric Utilization Systems, Latest Edition. Washington, DC: OSHA Department of Labor, U.S. Government Printing Office.

Photovoltaic Systems, Second Edition. James P. Dunlop. Orland Park, IL: American Technical Publishers.

Standard for Electrical Safety in the Workplace® (NFPA 70E®), Latest Edition. National Fire Protection Association (NFPA): Quincy, MA.

UL Standard 1703, UL Standard for Safety, Flat-Plate Photovoltaic Modules and Panels, Latest Edition. Camas, WA: Underwriters Laboratories.

UL Standard 1741, Standard for Inverters, Converters, Controllers and Interconnection System Equipment for Use with Distributed Energy Resources, Latest Edition. Camas, WA: Underwriters Laboratories.

Uniform Solar Energy Code, Latest Edition. Ontario, CA: International Association of Plumbing and Mechanical Officials (IAPMO).

Figure Credits

NCCER CURRICULA — USER UPDATE

NCCER makes every effort to keep its textbooks up-to-date and free of technical errors. We appreciate your help in this process. If you find an error, a typographical mistake, or an inaccuracy in NCCER's curricula, please fill out this form (or a photocopy), or complete the online form at **www.nccer.org/olf**. Be sure to include the exact module ID number, page number, a detailed description, and your recommended correction. Your input will be brought to the attention of the Authoring Team. Thank you for your assistance.

Instructors – If you have an idea for improving this textbook, or have found that additional materials were necessary to teach this module effectively, please let us know so that we may present your suggestions to the Authoring Team.

NCCER Product Development and Revision
13614 Progress Blvd., Alachua, FL 32615

Email: curriculum@nccer.org
Online: www.nccer.org/olf

❏ Trainee Guide   ❏ Lesson Plans   ❏ Exam   ❏ PowerPoints   Other _____

Craft / Level: _____   Copyright Date: _____

Module ID Number / Title: _____

Section Number(s): _____

Description: _____

_____

_____

_____

_____

Recommended Correction: _____

_____

_____

_____

_____

Your Name: _____

Address: _____

_____

Email: _____   Phone: _____

System Installation
and Inspection

57104-11

Trainees with successful module completions may be eligible for credentialing through NCCER's National Registry. To learn more, go to www.nccer.org or contact us at **1.888.622.3720.** Our website has information on the latest product releases and training, as well as online versions of our *Cornerstone* magazine and Pearson's product catalog.

Your feedback is welcome. You may email your comments to **curriculum@nccer.org,** send general comments and inquiries to **info@nccer.org,** or use the User Update form at the back of this module.

Objectives

When you have completed this module, you will be able to do the following:

1. Review the site assessment report, system design documents, and permits, and inspect the installation site.
2. Perform a job safety analysis (JSA) and deploy safety systems as needed.
3. Use system drawings and manufacturer's instructions to plan the installation and to inventory the project materials and tools needed for the job.
4. Locate structural members and install mounting hardware and raceway.
5. Inspect photovoltaic (PV) system components prior to installation.
6. Install the mechanical parts of the PV modules (panels) and balance-of-system components.
7. Install, label, and terminate electrical wiring and devices in accordance with local and national codes.
8. Activate and test the system to verify overall system operation.

Performance Task

Under the supervision of the instructor, you should be able to do the following:

1. Install and commission a system.

Prerequisites

Before you begin this module, it is recommended that you successfully complete *Core Curriculum* and *Solar Photovoltaic Systems Installer,* Modules 57101-11 and 57102-11. It is also suggested that you shall have successfully completed the following modules from the Electrical curriculum: *Electrical Level One,* Modules 26101 through 26111; *Electrical Level Two,* Modules 26201, 26205, 26206, and 26208 through 26211; *Electrical Level Three,* Modules 26301 and 26302; and *Electrical Level Four,* Modules 26403 and 26413.

Note: *NFPA 70®, National Electrical Code®,* and *NEC®* are registered trademarks of the National Fire Protection Association, Inc., Quincy, MA 02269. All *National Electrical Code®* and *NEC®* references in this module refer to the 2011 edition of the *National Electrical Code®.*

Contents

Topics to be presented in this module include:

Figures and Tables

Figures and Tables (*continued*) ────────

1.0.0 INTRODUCTION

After the site assessment and system design work is completed the final step is to install and activate the solar photovoltaic (PV) system. The level of safety concerns goes much higher because the work will be done on a roof or side of a building, or out in a field. If the installation is on a roof, there is a risk of damaging the customer's roof. Since the installation is dealing with components that generate, transmit, and control voltages, there is always the risk of electrical shock and/or arc flash. The wiring and terminations must be done properly for the overall system to work as designed, and to meet all applicable electrical and building codes. Only qualified electricians familiar with the *National Electrical Code®* (*NEC®*) articles that apply to PV systems are allowed to do any wiring on the PV components. A good installation is confirmed when the system is activated and everything works as planned.

2.0.0 JOB PREPARATIONS

Before any installation work is done, installers need to review the site assessment report and the system design plans that have already been done. Reviewing the documents prior to going onto the site allows the installer to see which materials and tools are needed for the job. If any of the materials or tools are not available, the installation work may need to be delayed until all the materials or tools are available. If something is missing and can be worked around, the work may continue until the missing materials or tools arrive. Such decisions need to be left to the installer's supervisor. Some of the most important things to review are the building permits.

2.1.0 Reviewing Site Assessment Reports

The installer needs to review the site assessment information to get an idea of what the conditions are at the site, and what must be done. This module focuses on an installation at a residential location. The following is a list of things to look for in the site assessment report:

- A sketch and any pictures of the installation site
- Ease of access to the installation site
- Height of installation site eaves
- Identified safety hazards or concerns
- Size of system being installed
- Estimated square footage needed for installation

- Age, type, and condition of installation site (if a roof)
- Slope of mounting surface.
- Type of mounting to be used (integral, flush, or standoff on a structure; rack on the ground; rack on a stationary pole; or rack on a tracking pole)
- Proposed location of panel mounts and raceways
- Proposed location of grounding electrodes and lightning arrestors
- Proposed location of balance-of-system (BOS) components:
  - Termination boxes
  - DC disconnects
  - Combiners
  - Inverter(s)
  - AC disconnects

Figure 1 shows the components of a typical solar panel system.

Based on data collected from the site assessment, the installer should have a clear idea of what he or she is getting into before going to the installation site. Arrangements for special equipment, such as an aerial lift, should be made prior to going onto the site. If scaffolding will be

On Site

Panel Adapter

The Safety Hoist Company has developed an adapter specifically designed to work with solar panels. The adapter is installed on a ladder hoist used by roofers, and the panels are placed onto the adapter so that they can be safely lifted to a roof. When the hoist reaches the top of the ladder hoist, the adapter can be tilted so that the panel(s) can be safely off-loaded.

104SA01.EPS

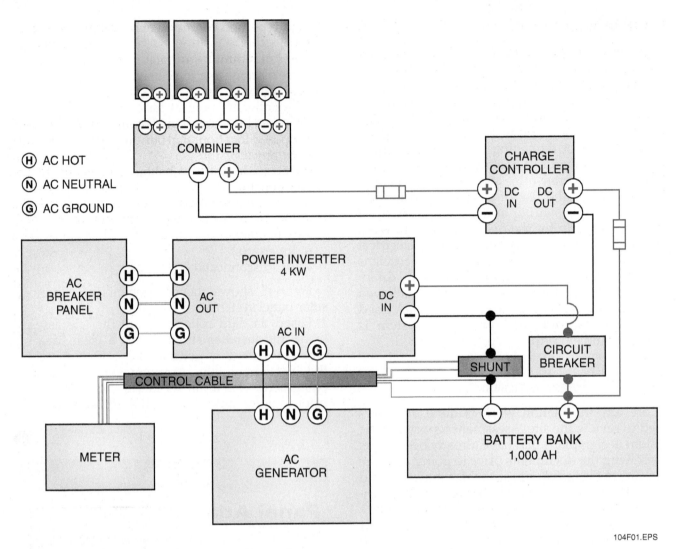

H AC HOT
N AC NEUTRAL
G AC GROUND

COMBINER

CHARGE CONTROLLER
DC IN   DC OUT

POWER INVERTER
4 KW

AC BREAKER PANEL
H H
N N   AC OUT
G G

AC IN
H N G

DC IN

SHUNT

CIRCUIT BREAKER

CONTROL CABLE

METER

AC GENERATOR
H N G

BATTERY BANK
1,000 AH

104F01.EPS

Figure 1 Drawing of a typical low-power solar panel system.

needed, that needs to be obtained and set prior to the installation crew arriving on site. Scaffold construction requires special training and certification. If an aerial lift or scaffolding is needed, there must be enough room around the base of the installation site to move the aerial lift or set the scaffolding. Plans also need to be made for at least one material hoist for lifting the mounting materials and panels onto the installation site. *Figure 2* shows a ladder hoist.

After all the site assessment materials have been reviewed, the next step is to review the system design documents for the specifics.

2.2.0 Reviewing System Design Plans

The site assessment report identified the customer needs and gave a rough estimate of what PV components would be needed. It also identified where they could be placed. The system

104F02.EPS

Figure 2 Ladder hoist.

design plan gives the specifics about the job. When the example site assessment for this module was initially done, the plan was to put a solar panel system onto the roof of a 1,500 square foot ranch-style house. *Figure 3* shows a sketch of the house and yard. The back roof of the house faces due south. It has full exposure to the sun with a solar irradiance rating of 5.17 kWh/square meter/day. The roof slopes at 20 degrees. The southern roof measures roughly 22' from gutter to ridgeline, and is 72' long east to west. There is a deck on the rear of the house. A chain-link fence surrounds the backyard, but it has a vehicle gate. There is an attic for accessing the underside of the roof. For this example, the panels will be installed in a standoff style of mounting. The plan is to install the panels in the area shown on the sketch.

This site gets some snow and ice, but not often. Snowfalls rarely exceed 5" and are usually gone in a few days. Normal winds range from 8 to 10 miles per hour (mph). Occasional gusts may be 30 mph or more, but those are rare.

2.2.1 Space for Panels

The 20-year old house has a new roof and uses an average of 1,632 kWh per month. The customer wants a system that will deliver 50 percent of the monthly power usage. According to the site assessment calculations, the customer would need a 6.55 kW system. The amount of needed roof space, in square feet, depends upon which type of panels are selected. If 100W panels were used, it would take roughly 66 panels to supply the customer's 6.55 kW system. The physical size of the panels varies between manufacturers and the amount of wattage they produce. An average 100W panel measures roughly 60" by 26" and takes up roughly 10 square feet. If 66 of those

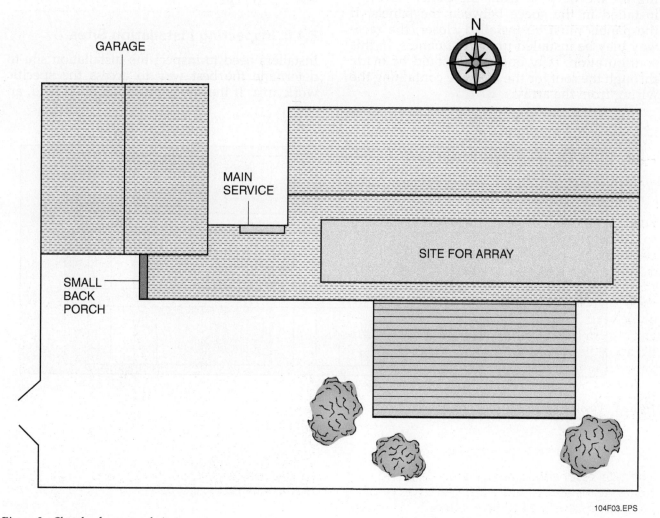

104F03.EPS

Figure 3   Sketch of proposed site.

100W panels were installed, they would take up a little more than 660 square feet. *Figure 4* shows a possible alignment of panels to form an array consisting of 66 panels.

The southern roof of the example house measured 70' long and 22' from gutter to ridgeline. It should have enough room for 60 or more panels.

Some building codes require that a space be left around the outer edges of an array and between the panels. The spaces are intended for emergency workers (fire fighters) who may have to move around on the roof. Maintenance workers could also use the space when work is needed on the panels. In the example shown, a 26" space was left between each row of panels.

The termination boxes on the panels will feed into combiners. In the configuration example, four panels will be fed into a combiner. The output of each combiner will be passed through the wiring in the raceway to the BOS components downstairs. The metal raceway connecting all the outputs from the panels is often installed in the space between the panels. If the panels must be installed closer, the raceway may be installed under the panels. In this configuration, only one entry would be made through the roof for the raceway containing the wiring from the array.

2.2.2 *Space for BOS Components*

The wiring from the panels will be fed to the west toward the garage. A bank of batteries, the charger controllers, the inverters, and the AC disconnects will be put in the garage. The eastern wall of the garage is approximately 15' from the main electrical service into the house. There is a 3' × 3' crawl space opening from the attic over the house into the attic over the garage. Raceway can be run from the house attic into the garage attic. From there, it can drop down into the garage to the BOS components.

2.3.0 Inspecting Building Permits

Before any work starts, all agreements with the owner, the utility, and any permitting agencies (such as city or county building inspectors) must be approved and signed. Most permitting agencies require that project drawings be submitted with the application. All the permits must be posted on the site until the work is signed off. Check the policies that apply to each installation site.

2.4.0 Inspecting Installation Sites

Installers need to inspect the installation site to determine the best way to access the specific work area. If the site is out in an open field, an

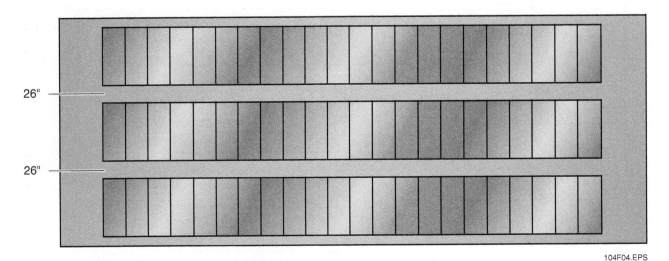

Figure 4 Possible installation of panels on a roof.

104F04.EPS

inspection needs to verify that the ground does not contain any kind of sewer line, septic line, gas line, water line, cable company or telephone company lines, or underground electrical lines. If such lines exist in the installation area, they must be clearly marked. If the installation site is on the roof of a building or home, the same lines need to be clearly identified around the structure. Aerial lifts or trucks hauling in the materials can damage these lines if they run over the ground above the lines.

Assuming that the installation site is on the roof of a residence, the installer needs to check the condition of the roofing material and the rafters, trusses, or joists under the roofing. Review the site assessment data to identify where the panels will start and stop on the roof. Roof-mounted solar arrays must be mounted directly to the roofing rafters, trusses, or joists. In some situations, additional spans of 2" × 6" wood sections can be installed between the rafters or trusses (*Figure 5*). These extra spans provide additional anchorage for the solar panel mounting attachments. Follow the manufacturer's requirements for anchor points and spacing.

After the exact installation points are identified, the installer can find, inspect, and mark the rafters, trusses, or joists that will be used to support and anchor the new panels. Most residential structures have wooden trusses and rafters. A strong stud finder can locate nails in the roofing materials from the outside, but cannot confirm that the nails are in the rafters. A visual inspection from inside the attic is needed to confirm rafter locations. *Figure 6* shows the building parts found in most residential structures.

If the building's roof has been built with metal roofing trusses or metal bar joists, the same inspections need to be made below the installation site. *Figure 7* shows examples of both metal roof trusses and bar joists.

> **WARNING!**
>
> While inspecting the installation site, keep an eye out for any electrical wires that may be run against, over and under, or near any of the rafters, trusses, or joists that will be used to anchor the array. Those wires need to be clearly identified and avoided.

While inside the attic or under an open roof checking the rafters, trusses, or joists, the installer should find and mark the truss, rafter, or joist nearest the start of the installation location. Identifying any rafter, truss, or joist from the outside is not easy. The installer may be able to locate a rafter or truss from the soffit and fascia area of the roof, but those areas only indicated where the lower ends of the rafters or trusses are located. The higher points on the rafters, trusses, or joists are much more difficult to locate. One option is to take a drill with a long, narrow-diameter bit into the area underneath the roof, and drill up through the structural member nearest the installation start point. This action can be done while inspecting the underside of the trusses or rafters. A hole can be drilled near the lower end of the structural member, and another hole can be drilled at the higher end of the member. Someone on the outside can mark each of the drilled holes with a marker. Drilling such holes in the selected structural member gives the installer high and low reference points from which all other measurements can be taken. *Figure 8* shows two reference points in a roof.

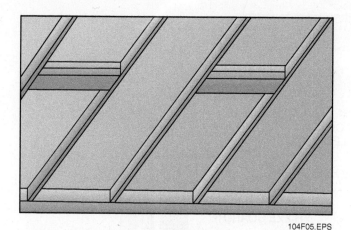

104F05.EPS

Figure 5   Spans between trusses.

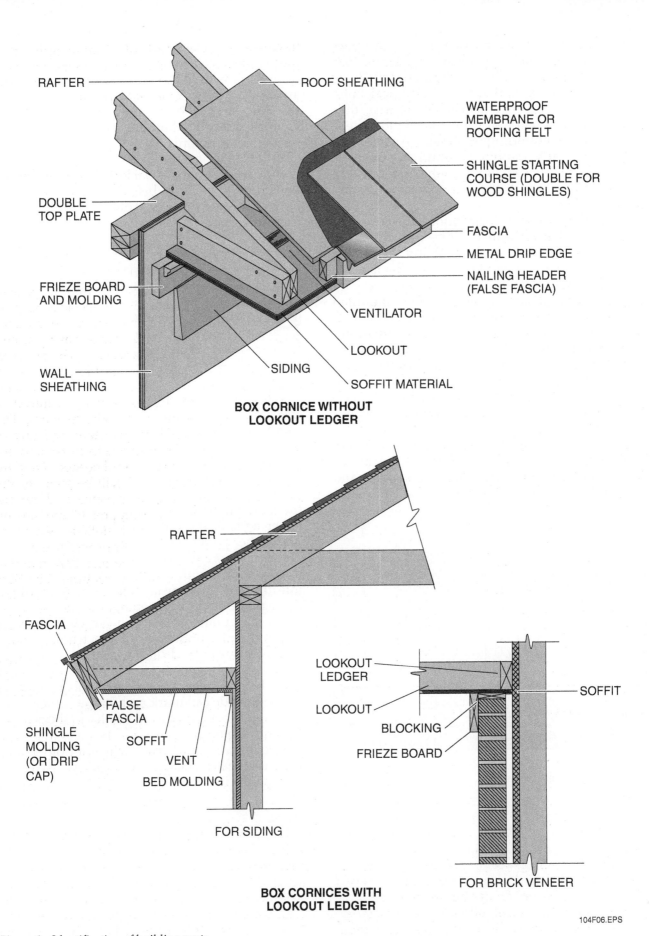

RAFTER

ROOF SHEATHING

WATERPROOF MEMBRANE OR ROOFING FELT

SHINGLE STARTING COURSE (DOUBLE FOR WOOD SHINGLES)

FASCIA

METAL DRIP EDGE

NAILING HEADER (FALSE FASCIA)

DOUBLE TOP PLATE

FRIEZE BOARD AND MOLDING

VENTILATOR

LOOKOUT

WALL SHEATHING

SIDING

SOFFIT MATERIAL

BOX CORNICE WITHOUT LOOKOUT LEDGER

RAFTER

FASCIA

LOOKOUT LEDGER

LOOKOUT

SOFFIT

FALSE FASCIA

SHINGLE MOLDING (OR DRIP CAP)

SOFFIT

VENT

BED MOLDING

BLOCKING

FRIEZE BOARD

FOR SIDING

FOR BRICK VENEER

BOX CORNICES WITH LOOKOUT LEDGER

104F06.EPS

Figure 6 Identification of building parts.

104F07A.EPS

104F07B.EPS

Figure 7   Metal trusses and bar joists.

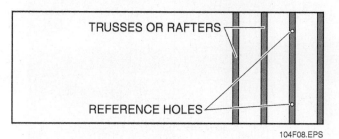

104F08.EPS

Figure 8   Reference points on roof.

3.0.0 SAFETY

A PV installation site, like any construction site, can be a dangerous place. An installation site out in a field is, or should be, much less dangerous than one on a structure. If the field installation is a matter of erecting the mounting hardware on the ground and installing the panels, those actions should be fairly safe. Things get much more complicated when the installation is on top of a structure such as a house or a building. Regardless of location, someone should do, or should have

already done, an analysis of all safety issues that apply to a given work site. The following are some generic job safety analysis issues that need to be addressed on any PV installation site:

- Housekeeping
- Fall protection
- Containment
- Personal protective equipment (PPE), including heat and sun protection
- Material lifting
- Batteries
- Electrical safety

3.1.0 Housekeeping

Before work begins on any site, the work area needs to be cleared of as many obstacles as possible. When the building materials are brought onto the site, they need to be laid out in a planned pattern. The mounting hardware will be used first. It should be stored in a location that is easy to reach and close to the installation site. The solar panels and the control devices (charge controllers and inverters) are the most delicate parts. The panels, charge controllers, inverters, and any of the other BOS components need to be secured out of the immediate work area until needed. They are expensive components, and will be some of the last items installed. The solar panels and mounting hardware items are shipped to the site on pallets. All materials to be installed need to be inventoried and stored in their proper places.

All of the materials for the installation arrive in some form of protective packaging. The BOS components need to be left in their shipping boxes until the components are ready to be installed. Packaging materials need to be contained and properly disposed of after they are removed. If the mounting hardware is not labeled after the packing materials are removed, some form of temporary labeling needs to be used to ensure that the parts are easily recognizable. After the packaging materials have been removed from the mounting hardware, the hardware items need to be neatly laid out for installation. They will most likely have protective shipping materials surrounding them. They need to stay on their pallets and in their protective materials until they are needed.

If the installation site is out in a field, there will possibly be digging tools in the area. Make sure that all tools are kept out of the walking areas. Exercise caution when walking around a field installation site. The ground may be uneven, which may cause either a tripping accident or a loss of balance. If holes have been dug for foundations, make sure that they are clearly marked with some kind of barrier. *Figure 9* shows different types of barriers and barricades that may be used around construction sites.

If the installation site is on the top or side of a structure, there will be ladders and possibly aerial lifts in the area. Keep walkways clear of tools and unused ladders. Be careful about walking near or under any overhead work. Also, be careful when working around moving machinery such as an aerial lift. Install barrier tape or barriers to keep others from getting too close to areas near working aerial lifts or any overhead work.

3.2.0 Fall Protection

There is always a risk of falling or dropping tools or equipment when either working on stand-alone solar installations or on the side or roof of a building. Anyone working in such environments must be careful to contain all tools and equipment within the immediate work area. Take only those tools or equipment items needed onto the roof or structure.

Some sites require the use of scaffolding. The scaffold guardrail system must be able to withstand a force of at least 200 pounds applied out

or down on the top rails. Scaffolds may only be erected and inspected by qualified individuals. Refer to *OSHA Standard 1926.502, Fall Protection Systems Criteria and Practices* for the specifics on scaffolding used for fall protection. Netting may be attached to scaffolding to contain falling objects.

Ladders used to gain access to a roof must extend a minimum of 36" above the surface of the roof. After getting onto a roof, installers must use some kind of anchor point that is strong enough to hold any attached load. The anchor point must be secured to roofing structures that can withstand a minimum of 5,000 pounds. Some newer roofs have fixed or permanent anchors. For those that do not have a fixed anchor, a temporary anchor may have to be installed. The Miller by Sperian company makes and sells some of the most reputable fall protection equipment in the industry. *Figure 10* shows two styles of roof anchors.

Installers must wear approved personal fall arrest system (PFAS) body harnesses with applicable lanyards. If a lanyard is attached to an anchor point below the D-ring on a harness, the worker will fall that distance plus the length of the lanyard. The increased force may cause more injury than if the anchor had been above the

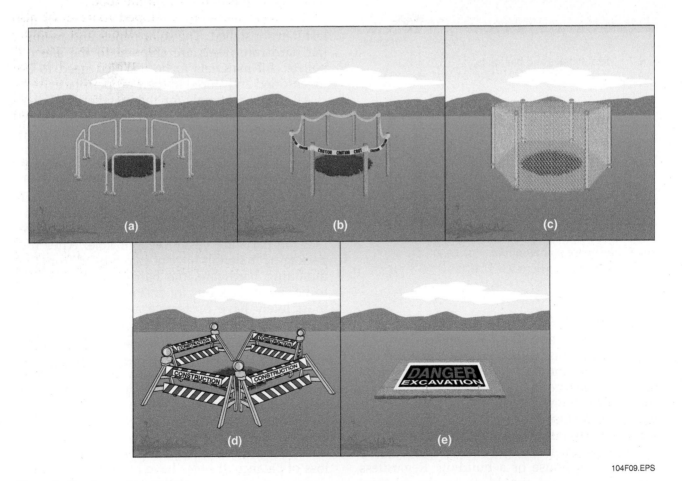

104F09.EPS

Figure 9 Barriers and barricades.

NCCER — *Contren® Learning Series* 57104-11

TEMPORARY ANCHOR WITH
SELF-RETRACTING LIFELINE

FIXED ANCHOR WITH LANYARD AND
AN ADJUSTABLE LANYARD

104F10.EPS

Figure 10 Fall protection anchors.

D-ring. To combat this, the American National Standards Institute (ANSI) has changed the equipment manufacturing standards for fall protection. The new standard is commonly referred to as the 12' lanyard. The lanyard is still only 6' long, but the allowable deployment (extendable length) of the shock absorber has been increased. This lessens the force transferred to the body through the harness, reducing the likelihood of injury.

When assessing the required free length of a PFAS, always account for the deployment of the shock absorber. In addition, PFAS manufacturers require up to a 3' safety factor. Taking this safety factor into account, *Table 1* indicates the minimum distance required between any hazard/obstruction and the worker or anchor point.

Clearly, a typical single-story building may not provide enough height to use standard lanyards. Retractable lanyards that engage and stop falls in less than 2' are usually the best choice for fall protection when used below 20' or on any type of lift equipment.

3.3.0 Containment

Solar panels can weigh 30 pounds or more, but a strong gust of wind can rip one out of an installer's hands. Installers need to ensure that they maintain control over the panels as they are being installed. Do not take more panels onto a roof than will be installed as soon as they are up there. The panels should be taken up and installed as soon as possible. Under certain wind conditions, the panels can act as a wing. Netting around the edge of a roof, or on the nearby scaffolding, will help contain objects that slide off the roof.

Installers working on sloped roofs must also keep their tools from falling off the roof. Companies such as Miller also make short lanyards for containing tools. The tool lanyards are attached to the tool and to the user's arm or belt. Workers on the ground need to be alert for objects that may fall from overhead work areas. Hardhats and safety glasses are essential to protect against falling or flying objects.

Be sure to mark and secure the site to keep unauthorized personnel out of the work area.

3.4.0 PPE and Heat or Sun Protection

Personal protective equipment for installers is a bit different than that worn by people doing other types of construction work. In addition

Table 1 Required Free Lengths for Various Lanyards

Lanyard Type	Lanyard Length	Shock Absorber Length	Average Worker Height	Safety Factor	Required Free Length
Standard 6' lanyard with 3.5' shock absorber	6'	3.5'	6'	3'	18.5'
New ANSI 6' lanyard with 4' shock absorber	6'	4'	6'	3'	19'
New ANSI 12' lanyard with 5' shock absorber	6' (12' freefall)	5'	N/A (accounted for by lanyard)	3'	20'

to the normal hardhats and fall protection harnesses that installers must wear, they also need to wear SPF 30 rated clothing that will protect their skin against excessive sun exposure. If protective clothing is not an option, then an appropriate sunscreen lotion needs to be used on exposed skin. Safety sunglasses are also needed for eye protection. Gloves are needed when handling PV panels. Nonslip shoes are also required.

PV installers work in some extremely hot environments most of the year. Temperatures on a roof are much higher than temperatures at ground level. Attic work also is much hotter. Air movement in attics is usually minimal. Air-filtering respirators may be required for attic work. Prolonged work on a roof or inside a hot attic can quickly result in dehydration. Drink eight ounces of water for every fifteen minutes of work in the heat. Caffeinated drinks and alcohol must be avoided prior to and during work hours. Installers need to be aware of heat exhaustion and heat stroke indications. They must watch each other for such signs.

When new systems are being installed, a PV system installer may have to work with FLA batteries, which are often used in PV systems. FLA batteries are normally shipped dry. Acid is shipped separately and added on site to activate the batteries. If an installer must work with battery acids, the installer must wear acid-resistant clothing, gloves, goggles, and boots. Battery acids can cause severe injuries if allowed to contact human skin. Anyone working with battery acids must obtain and review the latest material safety data sheet (MSDS) for the acid. The MSDS will specify what PPE is needed. *Figure 11* shows the MSDS for FLA solar batteries sold by the Concorde Battery Corporation. The MSDS also lists the action to take in case of an acid spill.

> **WARNING!**
>
> Never install batteries in an attic. The extreme temperatures of an attic location will result in battery leakage and premature failure.

3.5.0 Material Handling

The individual hardware items that make up a typical residential solar panel system are not all that heavy by themselves. The storage batteries are probably the heaviest individual components. They weigh between 95 and 120 pounds. The inverters weigh in the 75 to 100 pound range. Batteries, charge controllers, and inverters are normally kept at ground level. Charge controllers usually weigh between 10 and 15 pounds. Combiners and junction boxes or termination boxes weigh about the same, but are usually installed near the panels. Panels will range from 30 to 45 pounds. Installers need to work as a team when handling most of the PV materials. Always use approved lifting methods when lifting any load.

PV materials are normally brought to the site on some type of delivery truck. A forklift should be used to move pallets of equipment from the truck to the designated laydown area. After the delivered materials are on the ground, they can be unpacked and sorted. As stated earlier, the batteries, charge controllers, and inverters are the last items installed. They need to be secured out of the way until needed. The mounting hardware for the panels is the first to go onto the roof. It needs to be staged near the material lifting device that carries the materials to the roof. Solar panels should never be carried up a ladder onto a roof. Roofers use a ladder conveyor to transport their shingles onto a roof. Some ladder conveyor manufacturers are building adapters for their ladder conveyors (*Figure 12*) that will also carry solar panels to a roof.

Solar panel shapes make them awkward to handle when there is even a slight breeze. Stronger winds increase handling problems. Be extremely careful when moving solar panels from the material lifting device onto the roof. Some ladder conveyors are designed so that a top section of the conveyor tilts slightly over the gutter or eave and toward the roof. Such a conveyor would keep the installer on the roof from being too close to the edge of the roof.

3.6.0 Batteries

Chemical reactions inside a battery creates electricity. Batteries are made up of cells. Batteries have positive cells and negative cells. The cells are made up of plates. The plates are fixed and separated. Some form of electrolyte surrounds the plates. The electrolyte conducts electron flow between the positive and negative cells. As the chemical process takes place inside a battery, hydrogen gas is created. Since hydrogen is lighter than air, it tends to rise. If the level of hydrogen in a battery becomes excessive, the battery can explode. Because of that, batteries must be stored and installed in ventilated areas. The *NEC®* states that all battery locations must have ventilation, but the *NEC®* does not specify exactly how that ventilation must be done. Venting hydrogen is much different than venting combustion air from

CONCORDE BATTERY
VALVE REGULATED
LEAD ACID BATTERY
MATERIAL SAFETY DATA SHEET

Hazard Rating

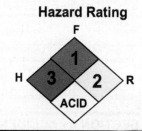

SECTION 1 – CHEMICAL PRODUCT AND COMPANY IDENTIFICATION

MANUFACTURER'S NAME:	CONCORDE BATTERY CORPORATION	EMERGENCY TELEPHONE NO.:	CHEMTEL 800-255-3924
ADDRESS:	2009 San Bernardino Rd., West Covina, CA 91790	OTHER INFORMATION CALLS:	626-813-1234
PERSON RESPONSIBLE FOR PREPARATION:	Gonzalo Ramos, Safety, Health & Environmental Affairs Manager	Revised Date:	OCTOBER 6, 2008

SECTION 2 - COMPOSITION/INFORMATION ON INGREDIENTS

C.A.S.	PRINCIPAL HAZARDOUS COMPONENT(S) (chemical & common name(s))	Hazard Category	% Weight	ACGIH TLV	OSHA PEL/TWA
7439-92-1	Lead/Lead Oxide (Litharge)/Lead Sulfate	Acute-Chronic	60-70	0.05 mg/m^3	0.05 mg/m^3
7440-70-2	Calcium	Reactive	<0.15	Not Established	Not Established
7440-31-5	Tin	Chronic	<1	2	2
7440-38-2	Arsenic (inorganic)	Acute-Chronic	<1	0.01	0.01
7664-93-9	Sulfuric Acid (Battery Electrolyte)	Reactive-Oxidizer Acute -Chronic	10-15	1.0	1.0

Note: PEL's for individual states may differ from OSHA's PEL's. Check with local authorities for the applicable state PEL's.
OSHA – Occupational Safety and Health Administration; ACGIH – American Conference of Governmental Industrial Hygienists; NIOSH – National Institute for Occupational Safety and Health.

COMMON NAME: (Used on label) Valve Regulated Lead-acid battery
(Trade Name & Synonyms) VRB, VRLA, SLAB, Recombinant Lead Acid: RG, GPL, AGM, PVX or FD Series, D8565 Series
Chemical Family: Toxic and Corrosive Material Mixture

Chemical Name: Battery, Storage, Lead Acid, Valve Regulated	Formula: Lead /Acid

SECTION 3 -- HAZARD IDENTIFICATION

Signs and Symptoms of Exposure	1. Acute Hazards	Do not open battery. Avoid contact with internal components. Internal components include lead and liquid electrolyte.
		Electrolyte - Electrolyte is corrosive and contact may cause skin irritation and chemical burns. Electrolyte causes severe irritation and burns of eyes, nose and throat. Ingestion can cause severe burns and vomiting.
		Lead - Direct skin or eye contact may cause local irritation. Inhalation or ingestion of lead dust or fumes may result in headache, nausea, vomiting, abdominal spasms, fatigue, sleep disturbances, weight loss, anemia and leg, arm and joint pain.
	2. Subchronic and Chronic Health Effects	Electrolyte - Repeated contact with sulfuric acid and battery electrolyte fluid may cause drying of the skin that may result in irritations, dermatitis, and skin burns. Repeated exposure to sulfuric acid mist may cause erosion of teeth, chronic eye irritation and / or chronic inflammation of the nose, throat, and lungs.
		Lead - Prolonged exposure may cause central nervous system damage, gastrointestinal disturbances, anemia, irritability, metallic taste, insomnia, wrist-drop, kidney dysfunction and reproductive system disturbances. Pregnant women should be protected from excessive exposure to prevent lead from crossing the placental barrier and causing infant neurological disorders.
		California Proposition 65 Warning: Battery posts, terminals, and related accessories contain lead and lead compounds, chemicals known to the State of California to cause cancer and reproductive harm, and during charging, strong
Medical Conditions Generally Aggravated by Exposure		Contact with internal components if battery is broken or opened, then persons with the following medical conditions must take precautions: pulmonary edema, bronchitis, emphysema, dental erosion and tracheobronchitis.
Routes of Entry	Inhalation - YES Ingestion – YES	Eye Contact- YES

Chemical(s) Listed as Carcinogen or potential Carcinogen	Proposition 65 - YES	National Toxicology Program - YES	I.A.R.C. Monographs - YES	OSHA - NO

SECTION 4 - FIRST AID MEASURES

Emergency and First Aid Procedures	Contact with internal components if battery is opened/broken.
1. Inhalation	Remove to fresh air and provide medical oxygen/CPR if needed. Obtain medical attention.
2. Eyes	Immediately flush with water for at least 15 minutes, hold eyelids open. Obtain medical attention.
3. Skin	Flush contacted area with large amounts of water for at least 15 minutes. Remove contaminated clothing and obtain medical attention if necessary.
4. Ingestion	Do not induce vomiting. If conscious drink large amounts of water/milk. Obtain medical attention. Never give anything by mouth to an unconscious person.

104F11A.EPS

Figure 11 MSDS for an FLA solar battery (1 of 3).

SECTION 5 - FIREFIGHTING MEASURES

Flash Point – Not Applicable	Flammable Limits in Air % by Volume: Not Applicable		Extinguishing Media – Class ABC, CO$_2$, Halon	Auto-Ignition Temperature	675°F (polypropylene)
Special Fire Fighting Procedures	Lead/acid batteries do not burn, or burn with difficulty. Do not use water on fires where molten metal is present. Extinguish fire with agent suitable for surrounding combustible materials. Cool exterior of battery if exposed to fire to prevent rupture. The acid mist and vapors generated by heat or fire are corrosive. Use NIOSH approved self-contained breathing apparatus (SCBA) and full protective equipment operated in positive-pressure mode.				
Unusual Fire and Explosion Hazards	Sulfuric acid vapors are generated upon overcharge and polypropylene case failure. Use adequate ventilation. Avoid open flames/sparks/other sources of ignition near battery.				

SECTION 6 - ACCIDENTAL RELEASE MEASURES

Procedures for Cleanup. Avoid contact with any spilled material. Contain spill, isolate hazard area, and deny entry. Limit site access to emergency responders. Neutralize with sodium bicarbonate, soda ash, lime or other neutralizing agent. Place battery in suitable container for disposal. Dispose of contaminated material in accordance with applicable local, state and federal regulations. Sodium bicarbonate, soda ash, sand, lime or other neutralizing agent should be kept on-site for spill remediation.

Personal Precautions: Acid resistant aprons, boots and protective clothing. ANSI approved safety glasses with side shields/face shield recommended.

Environmental Precautions: Lead and its compounds and sulfuric acid can pose a severe threat to the environment. Contamination of water, soil and air should be prevented.

SECTION 7 - HANDLING AND STORAGE

Precautions to be Taken in Handling and Storage	Store away from reactive materials, open flames and sources of ignition as defined in Section 10 – Stability and Reactivity Data. Store batteries in cool, dry, well-ventilated areas. Batteries should be stored under roof for protection against adverse weather conditions. Avoid damage to containers.
Other Precautions	GOOD PERSONAL HYGIENE AND WORK PRACTICES ARE MANDATORY. Refrain from eating, drinking or smoking in work areas. Thoroughly wash hands, face, neck and arms, before eating, drinking and smoking. Work clothes and equipment should remain in designated lead contaminated areas, and never taken home or laundered with personal clothing. Wash soiled clothing, work clothes and equipment before reuse.

SECTION 8 - EXPOSURE CONTROLS AND PERSONAL PROTECTION

Respiratory Protection (Specify Type)	None required under normal conditions. Acid/gas NIOSH approved respirator is required when the PEL is exceeded or employee experiences respiratory irritation.				
Ventilation	Store and handle in dry ventilated area.	Local Exhaust	When PEL is exceeded.	Mechanical (General)	Not Applicable
Protective Gloves	Wear rubber or plastic acid resistant gloves.		Eye Protection	ANSI approved safety glasses with side shields/face shield recommended	
Other Protective Clothing or Equipment	Safety shower and eyewash.				

SECTION 9 - PHYSICAL AND CHEMICAL PROPERTIES

Boiling Point: Not Applicable	Vapor Pressure	Not Applicable		Specific Gravity	1.250-1.320 pH <2	Melting Point: >320°F (polypropylene)	
Percent Volatile By Volume	Not Applicable	Vapor Density	Hydrogen: 0.069 (Air =1) Electrolyte: 3.4 @ STP (Air = 1)			Evaporation Rate	Not applicable
Solubility In water	100% soluble (electrolyte)			Reactivity in Water	Electrolyte – Water Reactive (1)		
Appearance and Odor:	Battery: Co-polymer polypropylene, solid; may be contained within an outer casing of aluminum or steel. Case has metal terminals. Lead: Gray, metallic, solid; brown/grey oxide Electrolyte: Odorless, liquid absorbed in glass mat material. No apparent odor.						

SECTION 10 - STABILITY AND REACTIVITY

Stability:	Stable	Conditions to Avoid: Avoid overcharging and smoking, or sparks near battery surface. High temperatures-cases decompose at >320°F.
Incompatibility (Materials to Avoid)		Sparks, open flames, keep battery away from strong oxidizers.
Hazardous Decomposition Products		Combustion can produce carbon dioxide and carbon monoxide.
Hazardous Polymerization		Hazardous Polymerization has not been reported.

SECTION 11 - TOXICOLOGICAL INFORMATION

GENERAL: The primary routes of exposure to lead are ingestion or inhalation of dust and fumes.

ACUTE:
INHALATION/INGESTION: Exposure to lead and its compounds may cause headache, nausea, vomiting, abdominal spasms, fatigue, sleep disturbances, weight loss, anemia, and pain in the legs, arms and joints. Kidney damage, as well as anemia, can occur from acute exposure.

CHRONIC:
INHALATION/INGESTION: Prolonged exposure to lead and its compounds may produce many of the symptoms of short-term exposure and may also cause central nervous system damage, gastrointestinal disturbances, anemia, and wrist drop. Symptoms of central nervous system damage include fatigue, headaches, tremors, hypertension, hallucination, convulsions and delirium. Kidney dysfunction and possible injury has also been associated with chronic lead poisoning. Chronic over-exposure to lead has been implicated as a causative agent for the impairment of male and female reproductive capacity, but there is at present, no substantiation of the implication. Pregnant women should be protected from excessive exposure. Lead can cross the placental barrier and unborn children may suffer neurological damage or developmental problems due to excessive lead exposure in pregnant women.

SECTION 12 - ECOLOGICAL INFORMATION

In most surface water and groundwater, lead forms compounds with anions such as hydroxides, carbonates, sulfates, and phosphates, and precipitates out of the water column. Lead may occur as sorbed ions or surface coatings on sediment mineral particles or may be carried in colloidal particles in surface water. Most lead is strongly retained in soil, resulting in little mobility. Lead may be immobilized by ion exchange with hydrous oxides or clays or by chelation with humic or fulvic acids in the soil. Lead (dissolved phase) is bioaccumulated by plants and animals, both aquatic and terrestrial.

104F11B.EPS

Figure 11   MSDS for an FLA solar battery (2 of 3).

SECTION 13 - DISPOSAL CONSIDERATIONS

Lead-acid batteries are completely recyclable. Return whole scrap batteries to distributor, manufacturer or lead smelter for recycling. For information on returning batteries to Concorde Battery for recycling call 626-813-1234. For neutralized spills, place residue in acid-resistant containers with sorbent material, sand or earth and dispose of in accordance with local, state and federal regulations for acid and lead compounds. Contact local and/or state environmental officials regarding disposal information.

SECTION 14 - TRANSPORT INFORMATION

All Concorde AGM, GPL, PVX, RG series and D8565 series are valve regulated lead acid (VRLA) batteries. Concorde's VRLA batteries have passed vibration, pressure differential and free flowing acid tests under CFR 49 173.159(d), meet IATA Special Provisions A48 & A67, and IMDG Special Provisions 238.1 & 238.2. The batteries are securely packaged, protected from short circuits and labeled "Non-Spillable." Concorde's VRLA batteries are exempt from DOT Hazardous Material Regulations, IATA Dangerous Goods Regulations, and IMDG Code.

US DOT
Excepted from the requirements because batteries have passed the vibration and pressure differential performance tests, and ruptured case test for Nonspillable designation.

IMO
Excepted from the requirements because batteries have passed the vibration and pressure differential performance tests, and ruptured case test for nonspillable designation. And, when packaged for transport, the terminals are protected from short circuit.

IATA
Excepted from the requirements because batteries have passed the vibration and pressure differential performance tests, and ruptured case test for nonspillable designation. And when packaged for transport, the terminals are protected from short circuit.

SECTION 15 - REGULATORY INFORMATION

U.S. HAZARDOUS UNDER HAZARD COMMUNICATION STANDARD:

LEAD - YES
ARSENIC – YES
SULFURIC ACID – YES

INGREDIENTS LISTED ON TSCA INVENTORY: YES

CERCLA SECTION 304 HAZARDOUS SUBSTANCES:

LEAD – YES   RQ: N/A*
ARSENIC – YES   RQ: 1 POUND
SULFURIC ACID – YES   RQ: 1000 POUNDS

* RQ: REPORTING NOT REQUIRED WHEN DIAMETER OF THE PIECES OF SOLID METAL RELEASED IS EQUAL TO OR EXCEEDS 100 μm (micrometers).

EPCRA SECTION 302 EXTREMELY HAZARDOUS SUBSTANCE:

SULFURIC ACID – YES

EPCRA SECTION 313 TOXIC RELEASE INVENTORY:

LEAD – CAS NO: 7439-92-1
ARSENIC – CAS NO: 7440-38-2
SULFURIC ACID – CAS NO: 7664-93-9

SECTION 16 - OTHER INFORMATION

THE INFORMATION ABOVE IS BELIEVED TO BE ACCURATE AND REPRESENTS THE BEST INFORMATION CURRENTLY AVAILABLE TO US. HOWEVER, CONCORDE BATTERY MAKES NO WARRANTY OF MERCHANTABILITY OR ANY OTHER WARRANTY, EXPRESSED OR IMPLIED, WITH RESPECT TO SUCH INFORMATION, AND WE ASSUME NO LIABILITY RESULTING FROM ITS USE. USERS SHOULD MAKE THEIR OWN INVESTIGATIONS TO DETERMINE THE SUITABILITY OF THE INFORMATION FOR THEIR PARTICULAR PURPOSES. ALTHOUGH REASONABLE PRECAUTIONS HAVE BEEN TAKEN IN THE PREPARATION OF THE DATA CONTAINED HEREIN, IT IS OFFERED SOLELY FOR YOUR INFORMATION, CONSIDERATION AND INVESTIGATION. THIS MATERIAL SAFETY DATA SHEET PROVIDES GUIDELINES FOR THE SAFE HANDLING AND USE OF THIS PRODUCT; IT DOES NOT AND CANNOT ADVISE ON ALL POSSIBLE SITUATIONS, THEREFORE, YOUR SPECIFIC USE OF THIS PRODUCT SHOULD BE EVALUATED TO DETERMINE IF ADDITIONAL PRECAUTIONS ARE REQUIRED.

The data/information contained herein has been reviewed and approved for general release on the basis that this document contains no export-controlled information.

FORM MSDS REV. 10/06/2008

104F11C.EPS

Figure 11   MSDS for an FLA solar battery (3 of 3).

Figure 12   Ladder conveyor.

appliances. A spark near or in the vented gas can create an explosion.

When a battery is completely discharged, it has zero volts across its positive and negative terminals. When an outside electrical supply is connected to the terminals, the cells charge up to a point when they can hold no more charge. Batteries that have been electrically charged will hold that charge until some electrical appliance or load is connected to the terminals. When the load is activated and starts to draw power, it reduces that charge. If allowed to continue, the load eventually discharges the battery to zero volts again. The battery used in a car is designed to charge and discharge quickly. The batteries used in solar energy system are designed to take much deeper charges. They are also designed to hold those charges much longer, and to discharge slowly. Deep cycle batteries are used for storing electrical energy created by wind generators and solar panels.

There are primarily three types of deep cycle batteries used in solar panel systems. They are flooded lead acid (FLA) batteries, sealed gel batteries, and AGM batteries. The gel and AGM batteries are part of a battery group known as valve-regulated lead acid (VRLA) batteries. Both types are sealed. FLA batteries have openings at the top. The AGM and sealed gel batteries are supposed to be maintenance free. They are also designed so that they do not have to be vented.

AGM batteries are considered to be the best of the group, but the gel and FLA batteries are the ones more often used to store solar energy. Due to their design, the gel and AGM batteries are much safer than FLA batteries because they do not vent any gases. *Figure 13* shows an FLA battery that is often used with solar panel systems.

The charging and discharging processes taking place inside a wet cell or FLA battery depletes the electrolyte inside the battery. Because of that, the FLA batteries must be refilled at times. They also must be vented. When FLA batteries are bought, they are usually transported and delivered dry. To activate them, the electrolyte solution must be poured into them. That is where it gets dangerous. The electrolyte is made with acid. Anyone working with acids must be extremely careful. Special PPE must be worn in case there is an acid spill or splash. Check the MSDS for the battery to see what PPE is needed for handling the battery or the acid in it.

3.6.1 Lifting and Leakage

Batteries are heavy, and lifting them can cause injuries. Once installed in their final storage location, they must be secured. If a battery is dropped or falls, it can do serious damage to whatever it hits. Sealed gel and AGM batteries do not present that much of a spill or leak threat if they are turned over because they should not leak. That

Figure 13   Typical FLA battery.

cannot be said for an FLA battery. Since the housing of an FLA battery must be vented, it can leak when turned over. If the fill caps on top of the battery are not securely fastened, they too can leak if the battery is turned over. The housing of an FLA battery is designed to contain the electrolyte and fumes as the battery operates, but if one of the threaded caps is loose, anything can escape, even when the battery is upright.

3.6.2 Placement and Venting

FLA batteries should never be placed directly under any electronic components. The corrosive vapors from these batteries can degrade and damage the circuits of the electronic components. FLA batteries should be stored inside vented enclosures that will safely contain any electrolyte that could be released if the battery case was to crack. Baking soda is a good product to keep near FLA batteries in case of a spill or leakage. The baking soda helps neutralize the acid. If nickel-cadmium batteries are being used, the acid base of the electrolyte is strong. Vinegar, which is a form of acetic acid, should be kept close to nickel-cadmium batteries in case of a spill or leakage.

3.6.3 Electrical Hazards

Electrical currents in excess of 10,000 amps can occur through a short circuit between the positive and negative terminals of a battery. If allowed to continue, the short circuit may burn open before the battery explodes. In either case, it is a dangerous situation. Because of such high current situations, batteries must have overcurrent protection when installed. That protection must have enough interrupting rating to end the short-circuit current situation. Extreme care must be used when installing batteries. Never allow tools or metal objects to fall across the terminals of a battery.

> **WARNING!**
> Remember that batteries must always be considered energized. Never ground or short a battery.

3.6.4 Environmental Considerations

One of the most important considerations for battery installation is the environment in which they are placed. Batteries do not perform well in extreme temperatures. Ideally, they should be placed in an environment with a constant temperature of 77°F. Refer to the manufacturer's manual for temperature specifics. The electrolyte in a battery freezes in extremely cold temperatures. If a battery freezes, let it thaw before trying to charge it. When trying to thaw a frozen battery, place it in an isolated area where it can do the least amount of damage if its case explodes.

Another environmental consideration is what to do with leftover acids or used batteries. Most states have specific guidelines about how to dispose of batteries and acids. Check with the local Department of Health for specific guidelines. The officials at a local landfill should have the specific guidelines.

3.6.5 Wiring

In theory, all the batteries used in a battery bank should be the same type with the same charge and discharge characteristics. To minimize any voltage drops, large (2/0 or 4/0) cabling should be used between the batteries. The batteries may be wired in series or in parallel, depending upon what the designer wants. Installers need to make sure that they clearly understand which battery cables to use, and how the batteries should be connected. Improper connections can damage the batteries and could cause an explosion. The *NEC®* permits, but does not require, that flexible cables be used for connecting batteries, but such cables help reduce the stress on the battery terminals. Welding cables, automotive battery cables, and other such cables are not *NEC®* acceptable for connecting solar storage batteries. When interconnecting batteries, all jumpers must be the same length and wire gauge.

3.7.0 Electrical Safety

NEC Section 110.26 specifies the minimum amount of working spaces needed to operate equipment safely. Review those specifics before beginning any PV installation job. Working spaces are especially important when it comes to installing the BOS components. A clearance of 3' is a generally accepted distance for most PV components. Greater distances are required with higher (150V to 600V) levels of voltage. For DC voltages less than 60V, smaller working spaces may be permitted under special circumstances. Check local codes for specifics.

Installers, and anyone else involved in the installation of a PV system, must keep in mind that all this PV equipment is designed to create, transmit, store, and deliver electrical power. The following are the primary electrical safety concern areas on a PV system:

- Grounding
- Panels and their outputs
- Charge controllers and DC disconnects
- Batteries
- Inverters and AC disconnects

Figure 14 shows a diagram of a 4 kW solar system with the various components identified.

3.7.1 Grounding

Since the purpose of PV systems is to produce electricity, grounding for the PV components is a must. The grounding rod (electrode) used for an existing AC service can also be used for grounding the inverter(s). For additional lightning protection, the PV array (mounting hardware and panels) needs to be tied directly to its own grounding electrode. *Figure 15* shows grounding options for PV system components.

Grounding for PV systems is covered in *NEC Article 690, Part V*. If the maximum voltage produced by a PV system exceeds 50V, then one conductor must normally be grounded. Check *NEC Section 690.35* for details on how to install ungrounded PV systems of any voltage. There are two types of grounding associated with PV systems. They are equipment grounding and system grounding.

Equipment grounding is intended to prevent electrical shocks from the conductive (metal)

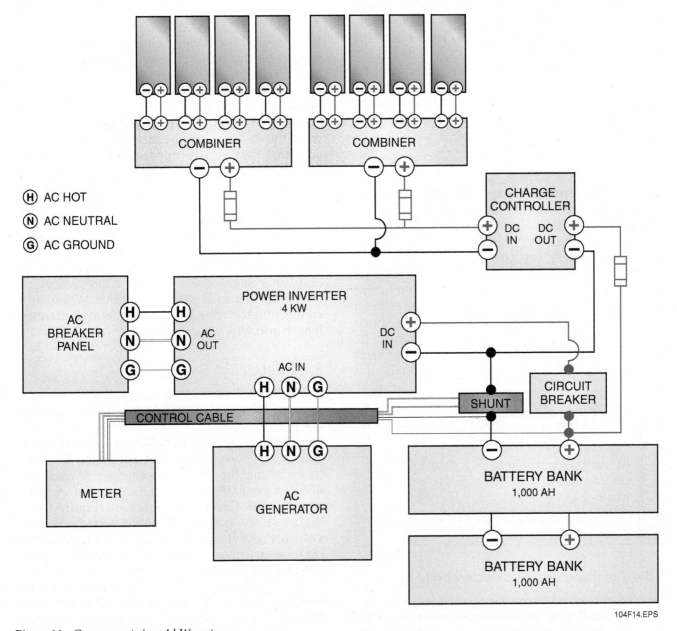

Figure 14 Components in a 4 kW system.

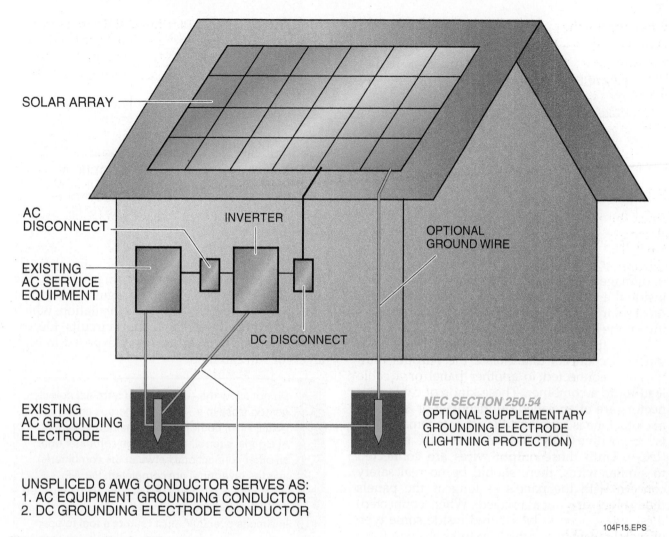

SOLAR ARRAY

AC DISCONNECT

INVERTER

EXISTING AC SERVICE EQUIPMENT

OPTIONAL GROUND WIRE

DC DISCONNECT

EXISTING AC GROUNDING ELECTRODE

NEC SECTION 250.54
OPTIONAL SUPPLEMENTARY GROUNDING ELECTRODE (LIGHTNING PROTECTION)

UNSPLICED 6 AWG CONDUCTOR SERVES AS:
1. AC EQUIPMENT GROUNDING CONDUCTOR
2. DC GROUNDING ELECTRODE CONDUCTOR

104F15.EPS

Figure 15   Grounding for PV system.

parts that do not normally carry electrical current. Mounting frames, metal raceway, and junction or terminal boxes are examples of equipment that needs to be grounded. If a live electrical wire touches an ungrounded metal part, such as those just listed, the object becomes electrified. The result is a ground fault. If a person comes in contact with the electrified object, the human body provides a path to ground and thus becomes shocked. If the shock is strong enough, that person will be injured and may be killed. Installing grounds on such equipment reduces or eliminates the risk of electrical shock from the equipment items. Equipment grounding wires are either green, green with yellow stripes, or they are bare.

System grounding takes one current-carrying conductor and attaches it to ground. On DC voltage systems, that grounded leg can be either the negative or the positive conductor (white or gray wire), depending upon how the PV panels and inverters are set up. In AC voltage systems, the

neutral conductor can be either white or gray wire. The closer the grounding wire can be connected to the PV source (the panels), the better the lightning protection will be. *Figure 16* shows different grounds and disconnects on a PV system.

3.7.2 Panels and Their Outputs

The electrical power from the solar panels is direct current (DC) power. It can be deadly under the right conditions. The solar panels become active as soon as they are exposed to sunlight. Some manufacturers and installers recommend that the panels be kept under some kind of cover until they are wired in. Keeping them covered should reduce the risk of the panels generating any electricity.

The way the wiring inside a panel is tied into a junction box, or termination box, the cells and modules on the panel should pose no threat unless the materials surrounding them are broken

System Installation and Inspection

17

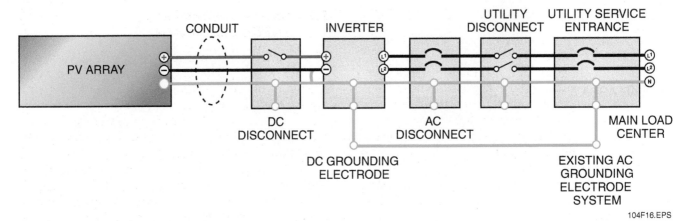

Figure 16 PV system grounding and disconnects.

or damaged. Those wires attach to terminal points inside the terminal box found on the backside of most solar panels. *Figure 17* shows the output wiring on the back of a solar panel.

The output wires from the terminal boxes have quick disconnects that isolate the outputs until they are connected to another panel or a cable leading to a combiner. These kinds of cable connectors are often called multicontact or MC connectors. One is male and the other is female. One is DC positive (DC+) and the other is DC negative (DC–). Until those output wires are connected to similar wires, there should be no real safety concern with the panels as long as the panels themselves are not damaged. When connected, these wires need to be secured inside some type of metal raceway or cable clip to keep them from

flapping in the wind or rubbing against the mounting hardware. Rubbing against electrical wiring usually breaks down the insulation, which leads to exposed wires and short circuits. Electrical wiring inside metal raceway exposed to high levels of sun can melt.

> **WARNING!**
>
> Do not allow the + lead from a terminal box get too close to a – lead. Dangerous arcing can result. Also, do not connect the output wires of a panel's terminal box to the combiner until all other connections between the combiner(s) and the inverter(s) are completed. Per *NEC Section 690.33(C) and (E)*, all connectors must be of the locking and latching type. Connectors used in circuits over 30V must require a tool to open.

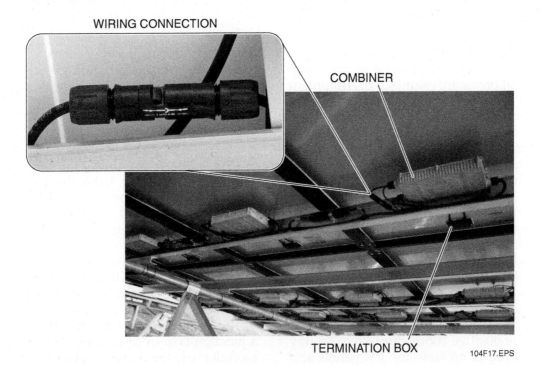

Figure 17 Solar panel outputs.

3.7.3 Combiners

The combiners used with PV systems are nothing more than terminal panels. Some have fused circuits and other do not. Some combiners have built-in disconnects that allow the panel inputs to be disconnected from the rest of the PV circuits. Combiners range in size from a two-input combiner to one with multiple inputs from different solar panels. The incoming wires may be from single panels or a string of panels tied together. The wires feed into one side of the combiner's terminal board where the voltages are combined onto a single output line. *Figure 18* shows the terminals inside of a typical combiner.

This particular combiner can receive inputs from six strings of PV panels. The positive inputs pass through fuses and combine at the positive

> **WARNING!**
>
> Be extremely careful when connecting panel wires because they may be carrying voltage from the panels. Only qualified personnel may make electrical connections.

busbar. The DC+ output of the fuse block exits at the fused output to a charge controller or an inverter. The nonfused output from the negative busbar of the combiner is also tied to the same devices. Remember that the input wire from each string of panels may represent up to 48 VDC. Electrically these six inputs represent six parallel circuits with 48 VDC on each. The voltage is not all that high, but the current may be.

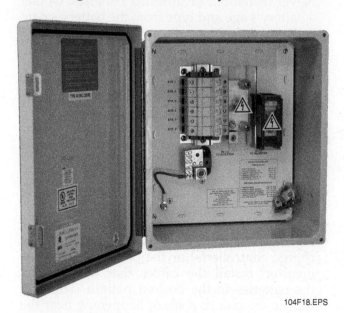

104F18.EPS

Figure 18 Combiner.

3.7.4 Charge Controllers and DC Disconnects

Charge controllers are only used in a PV system if batteries are used. Solar batteries can be overcharged if allowed to stay connected to the supplying source too long. A charge controller monitors the charge on a battery, and then controls how the battery is charged. The charging process may vary from a trickle charge to a surge. The control devices inside the charge controller make those decisions. When the batteries are fully charged, the charge controller diverts the solar panel power past the batteries and directly into the inverter(s). *Figure 19* shows a simplified diagram of the wiring into and out of a charge controller.

From a safety point of view, charge controllers are essentially sealed units. They will have input wires from the combiners. Their output wires go to the batteries or an inverter. Ground-fault protection devices are normally connected at the input of the charge controllers. There should also be some kind of output DC disconnect switch between the charge controller and the battery bank. There may be a separate DC disconnect switch between the controller and the inverter.

Before installing a charge controller, review the User Manual to make sure that the controller circuits can handle the incoming power from the PV panels. Controllers are designed to handle a maximum amount of power (voltage and current). The controller must be able to handle at least 125 percent of any short-circuit current produced by the PV panels. Refer to *NEC Section 690.7* for details. *NEC Sections 690.15 and 690.17* deal with the disconnects into and out of the controllers.

Before installing or working on any charge controller, make sure that the disconnects into and out of it are opened. Charge controllers have capacitors in them that retain electrical charges. Refer to the User Manual for details on how long it takes the capacitors to discharge after power is removed. If in doubt, allow at least five minutes for any capacitors to discharge. When sure that the charge controller is electrically safe, continue with the installation process.

3.7.5 Batteries

The batteries used in PV systems come in sealed or vented designs. Each design has different charging and discharging characteristics. Review the charging instructions in the manufacturer's instructions for the batteries in use.

Extremely high electrical currents can occur through a short circuit between the positive and

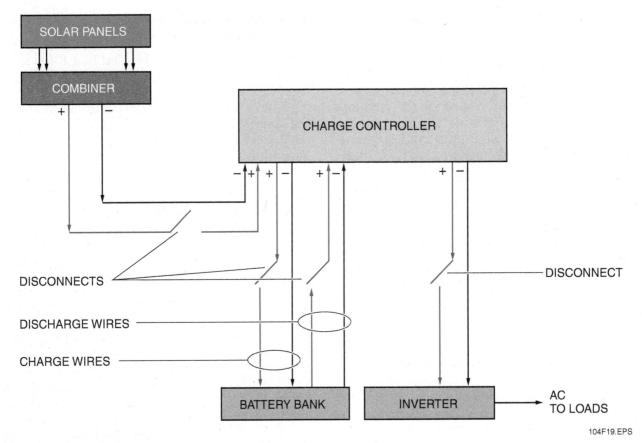

Figure 19 Basic charge controller wiring.

104F19.EPS

negative terminals of a battery. Batteries are usually shipped with plastic caps covering the positive terminals. Leave the terminal caps in place until the batteries are ready to be wired into the PV system.

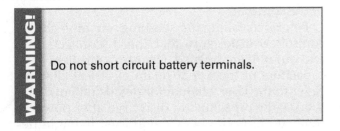

WARNING!

Do not short circuit battery terminals.

If there is any possible way for voltage from the batteries to back feed into a charger controller, make sure to disconnect the batteries from the controller before working on the controller. The same goes for anything else that may be receiving power from the batteries. The following are a few general safety instructions concerning batteries:

- Ensure that all electrical work is done in accordance with applicable electrical codes.
- Remove all jewelry prior to working on any electrical equipment.
- Wear all PPE required for battery work.

- Use insulated tools to reduce the risk of short circuits.
- Never work on batteries alone.
- Never smoke near batteries.
- Use proper lifting techniques when lifting or moving batteries.
- Ensure that all batteries are the same type. Never install old or untested batteries. Obtain a battery's age and type from the battery's date code or label.
- Read and understand all manufacturer instructions, cautions, and warnings about the batteries being installed or used.
- Install batteries with at least 1" of space between their cases for cooling and ventilation.
- Install batteries only in well-vented area. Vents go to the outside atmosphere. If installed inside enclosures, make sure that the enclosure is vented at the highest point.
- Review the battery alignment and connection plan before connecting any battery cables.
- Verify polarities, voltage levels, and cable sizing.
- Before connecting the power cable from the charge controller(s) or the power cable to the inverter, install the cables that will connect the batteries in the desired pattern. This will reduce the risk of a spark happening near the batteries.

- If battery electrolyte is spilled or splashed, wash down the area with soap and water.
- If battery acid gets into the eyes, flood them with running cold water for at least 15 minutes and seek medical attention immediately.
- Recycle old or unusable batteries.

3.7.6 Inverters and AC Disconnects

There are some inverters on the market that combine the functions of an inverter with those of a charge controller. Regardless of configuration, the result is the same. The DC power from the PV panels or their combiners is eventually fed into the input terminals of an inverter. The inverter changes the DC power into AC power. The inverter's AC output goes into a conventional circuit breaker (disconnect). The circuit breaker may be installed in the user's main service connection, or in a separate feed to some AC appliance. The following are a few general safety instructions for inverters:

- Ensure that all electrical work is done only by a licensed electrician.
- Ensure that all electrical work is done in accordance with applicable electrical codes.
- Remove all jewelry prior to working on any electrical equipment.
- Read and understand all manufacturer instructions, cautions, and warnings about an inverter before installing or using it.
- Identify all input and output terminals that may have live voltages on them. Inputs may be from batteries or the PV array.
- Use insulated tools to reduce the risk of short circuits.
- Mount inverters only in inside areas protected from liquids. Do not expose them to any liquids (rain, snow, or water from building washes).

4.0.0 JOB PLANNING AND INVENTORY OF MATERIALS AND TOOLS

After the site assessment and system design have been completed, and the PV system components have been selected, the installer should have a pretty good idea of what must be done. When a job has been properly planned, the installer should receive a copy of the planned installation paperwork, along with a copy of the drawings, and a Bill of Materials. In some cases, the Bill of Materials is printed on the drawings. An installer should be able to take a drawing and the related Bill of Materials, and visually go through the drawing and identify the components. That same Bill of Materials should be used then to inventory the hardware brought onto the installation site. *Figure 20A* shows an example of an electrical drawing with component information. *Figure 20B* shows an example of a general arrangement drawing with component and distance information.

Earlier in this module, an overhead view (sketch) of the proposed installation site was given, along with measurements of the roof area available for the installation. The site assessment calculated that the customer would need a 6.55 kW system to meet the targeted needs. After the system designer finished running the numbers and making the corrections to compensate for distances and loads, the final design called for an 8.0 kW system, which was going to take up more roof space than was planned for or available. Due to the larger size of the system, the designer also determined that multiple charge controllers and inverters would be needed. *Table 2* shows a chart that can be used to compile a list of materials that would be needed to install this system.

The PV system just described is much too large to address in this module. For training purposes, this module focuses on installing four panels in a standoff position that positions them between 4" and 6" above a shingled roof that slopes 20 degrees. Regardless of the size, an installation job can be broken down into the following three segments:

- Installation of mounting hardware
- Installation of panels and wiring to the balance-of-system components
- Installation and connection of the BOS components

4.1.0 Mounting Hardware

In an earlier section, the original plan was to install the array along the southern roof of the customer's house. After the system designer made the needed corrections, the system was changed into an 8 kW system, which is roughly 23 percent larger than originally planned. As a result, the type and number of panels needed had to be changed. The final plan came up with 33 panels by Canadian Solar, with each producing 240 watts of power. Each of those panels is 63.1" long, 41.8" wide, and 1.57" deep. Each weighs 44 pounds. *Figure 21* shows a Canadian Solar CS5P-M panel. The number of mounting hardware items will be determined by how these panels are installed.

4.1.1 Anchor Devices

Arrays mounted on roofs can be laid out in different directions. The mounting beams or rails can be laid parallel to the rafters, which is the easier alignment, or perpendicular to the rafters.

System Installation and Inspection

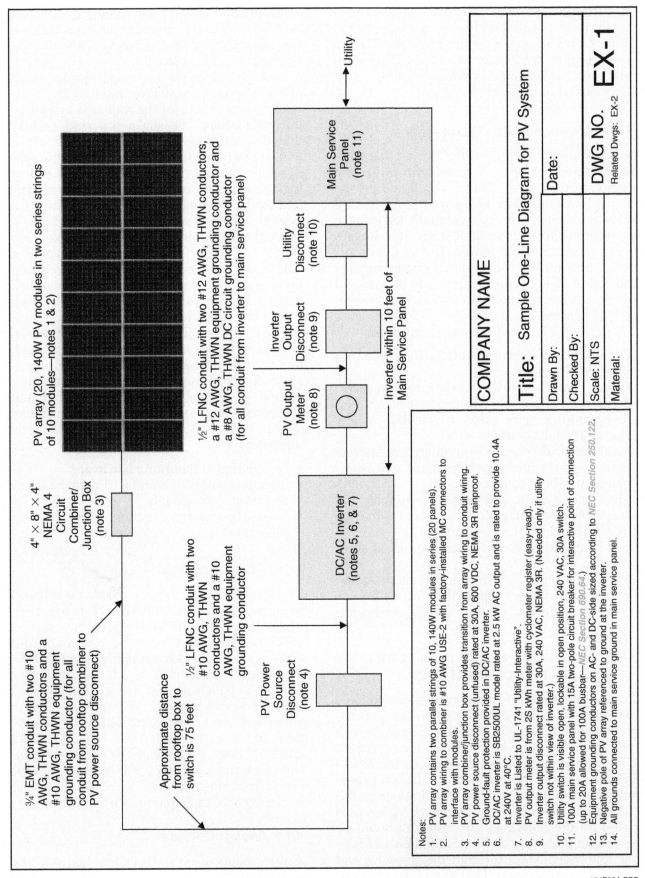

PV array (20, 140W PV modules in two series strings of 10 modules—notes 1 & 2)

3/4" EMT conduit with two #10 AWG, THWN conductors and a #10 AWG, THWN equipment grounding conductor (for all conduit from rooftop combiner to PV power source disconnect)

4" × 8" × 4" NEMA 4 Circuit Combiner/ Junction Box (note 3)

Approximate distance from rooftop box to switch is 75 feet

1/2" LFNC conduit with two #10 AWG, THWN conductors and a #10 AWG, THWN equipment grounding conductor

PV Power Source Disconnect (note 4)

DC/AC Inverter (notes 5, 6, & 7)

Inverter within 10 feet of Main Service Panel

1/2" LFNC conduit with two #12 AWG, THWN conductors, a #12 AWG, THWN equipment grounding conductor and a #8 AWG, THWN DC circuit grounding conductor (for all conduit from inverter to main service panel)

PV Output Meter (note 8)

Inverter Output Disconnect (note 9)

Utility Disconnect (note 10)

Main Service Panel (note 11)

Utility

Notes:
1. PV array contains two parallel strings of 10, 140W modules in series (20 panels).
2. PV array wiring to combiner is #10 AWG USE-2 with factory-installed MC connectors to interface with modules.
3. PV array combiner/junction box provides transition from array wiring to conduit wiring.
4. PV power source disconnect (unfused) rated at 30A, 600 VDC, NEMA 3R rainproof.
5. Ground-fault protection provided in DC/AC inverter.
6. DC/AC inverter is SB2500UL model rated at 2.5 kW AC output and is rated to provide 10.4A at 240V at 40°C.
7. Inverter is Listed to UL-1741 "Utility-Interactive".
8. PV output meter is from 2S kWh meter with cyclometer register (easy-read).
9. Inverter output disconnect rated at 30A, 240 VAC, NEMA 3R. (Needed only if utility switch not within view of inverter.)
10. Utility switch is visible open, lockable in open position, 240 VAC, 30A switch.
11. 100A main service panel with 15A two-pole circuit breaker for interactive point of connection (up to 20A allowed for 100A busbar—*NEC Section 690.64*).
12. Equipment grounding conductors on AC- and DC-side sized according to *NEC Section 250.122*.
13. Negative pole of PV array referenced to ground at the inverter.
14. All grounds connected to main service ground in main service panel.

COMPANY NAME

Title: Sample One-Line Diagram for PV System

Drawn By:

Checked By:

Scale: NTS

Material:

Date:

DWG NO. EX-1

Related Dwgs: EX-2

104F20A.EPS

Figure 20 (A) Electrical drawing with components identified.

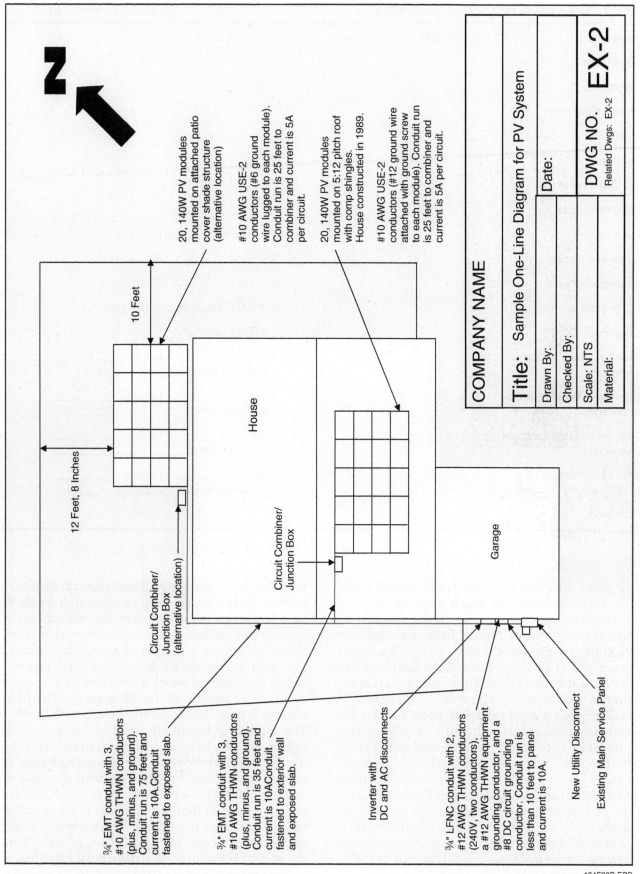

Figure 20 (B) General arrangement drawing with components and distances identified.

 System Installation and Inspection

Labels within the diagram:

20, 140W PV modules mounted on attached patio cover shade structure (alternative location)

#10 AWG USE-2 conductors (#6 ground wire lugged to each module). Conduit run is 25 feet to combiner and current is 5A per circuit.

20, 140W PV modules mounted on 5:12 pitch roof with comp shingles. House constructed in 1989.

#10 AWG USE-2 conductors (#12 ground wire attached with ground screw to each module). Conduit run is 25 feet to combiner and current is 5A per circuit.

COMPANY NAME

Title: Sample One-Line Diagram for PV System

DWG NO. EX-2

Related Dwgs: EX-2

Date:

Drawn By:

Checked By:

Scale: NTS

Material:

10 Feet

12 Feet, 8 Inches

House

Garage

Circuit Combiner/ Junction Box (alternative location)

Circuit Combiner/ Junction Box

¾" EMT conduit with 3, #10 AWG THWN conductors (plus, minus, and ground). Conduit run is 75 feet and current is 10A. Conduit fastened to exposed slab.

¾" EMT conduit with 3, #10 AWG THWN conductors (plus, minus, and ground). Conduit run is 35 feet and current is 10A. Conduit fastened to exterior wall and exposed slab.

Inverter with DC and AC disconnects

¾" LFNC conduit with 2, #12 AWG THWN conductors (240V, two conductors), a #12 AWG THWN equipment grounding conductor, and a #8 DC circuit grounding conductor. Conduit run is less than 10 feet to panel and current is 10A.

New Utility Disconnect

Existing Main Service Panel

104F20B.EPS

Table 2 Materials List for a Solar PV Installation

Items	Model	Quantity	Remarks
Solar Panels			
Combiners			
Charge Controllers			
Batteries			
Inverters			
DC Disconnects			
AC Disconnects			
Metal raceway			10' lengths of 3/4" EMT
90-degree elbows			3/4" EMT
45-degree elbows			3/4" EMT
Connectors			3/4" EMT (insulated and rain tight)
Couplings			3/4" EMT (rain tight)
Wire from array to charge controller			Type: _____ Length: _____ Installation rating: _____
Wire from charge controller to batteries			Type: _____ Length: _____ Installation rating: _____
Wire from charge controller to inverters			Type: _____ Length: _____ Installation rating: _____
Wire from inverters to AC breaker			Type: _____ Length: _____ Installation rating: _____
AC breaker			

Whichever direction is selected, the overall array needs to be centered as much as possible over the structural members of the roof. If unable to locate the individual rafters from the outside, have someone go into the attic and drill a small hole up and at an angle toward the center of the rafter closest to the edge of the installation area. Drilling a hole at the low end and the upper end of the rafter gives a good reference point to start the layout. The person inside the attic can also verify the distance between the rafters under the roof. After the reference holes are marked, a measuring tape, framing square, and a chalkline can be used to layout and mark the rest of the rafters. The outer perimeter of the array can also be marked with the chalkline. Panels must be installed with a little spacing between them so that the mounting hardware can be installed to hold the panels to the mounting beams.

To determine the overall footprint of an array, the installer must know the dimensions of the panels that will be used. After those measurements are known, add a 1" space between the modules for the middle clamps. Always follow the manufacturer's spacing instructions for the mounting system in use. When panels are arranged in a landscape orientation, add a 3" space for the end clamps. *Figure 22* shows examples of panel orientation.

There are specific pieces of mounting hardware used to secure the mounting beams to the rafters. The mounting hardware must be secured to the rafters, or to whatever structural items are being used to support the roof. The mounting hardware cannot be attached to only the sheeting under the shingles. Holes have to be drilled through the roofing materials so that lag (anchor) bolts or screws can be sunk into the rafters. Weatherproof sealer, applied with a caulking gun, is used to seal around the anchor bolts after they have been installed.

Mounting hardware varies widely by manufacturer. The environmental conditions in which the array is being built influences the type of mounting

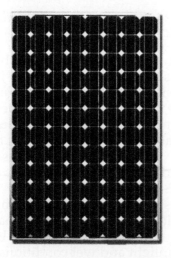

DIMENSIONS		
	METRIC	INCHES
LENGTH	1602 mm	63.1"
WIDTH	1061 mm	41.8"
THICKNESS	40 mm	1.57"
WEIGHT	20 kg	44 lbs

104F21.EPS

Figure 21 Typical 240W solar panel.

hardware used. Anchoring devices are designed to withstand specific wind shear and uplift forces. Inspectors may require a record of these ratings. In addition, many localities require that PV support structures be designed and approved by a professional engineer (PE). Check local requirements.

Figure 23 shows a beam-anchoring mount that is supplied with a section of waterproof flashing. To install this type of mounting device, the flashing is first slipped in under the shingles above the rafter to which the anchor will be attached. The flashing is then aligned so that the bolt hole in it is centered over the rafter. Next, a ¼" pilot hole for a ⅜" lag screw is drilled through the roofing and slightly into the rafter. Fill the hole with roofing caulk to ensure a weathertight seal. After the hole is drilled, the standoff block and flange connector are aligned over the hole. When all the parts are aligned, the anchor bolt is threaded down through all of them and the roofing materials into the rafter below. The bolt is tightened, but care must be taken to make sure that it does not strip out of the rafter.

If such anchors are used, two or more are used for each mounting beam. The beams fit down in between the fingers of the flange connectors. The number of anchors used depend upon the amount of loading and lifting stresses the designer expects to happen when the panels are installed.

LANDSCAPE

PORTRAIT

104F22.EPS

Figure 22 Landscape and portrait panel orientation.

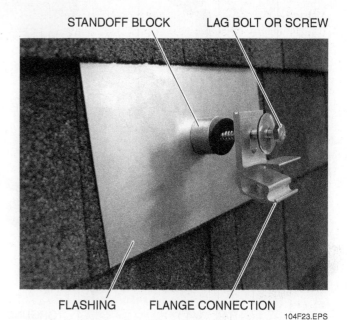

STANDOFF BLOCK   LAG BOLT OR SCREW

FLASHING   FLANGE CONNECTION

104F23.EPS

Figure 23 Beam-anchoring mount.

The standoff blocks determine how high the panels will eventually sit above the roof. When inventorying the materials, all the parts of such anchoring assemblies must be accounted for and kept secure until used. Anchoring devices, such as Quick Mount PV® assemblies, usually come as a package. If they are packaged, just count the packages.

4.1.2 Mounting Beams, Flange Connections, and Clamps

Different manufacturers build their mounting beams, anchoring devices, and panel clamps differently. This training module shows mounting hardware made by UNIRAC® and CLICKSYS™. *Figure 24* shows typical solar panel mounting hardware.

4.2.0 Panels, Combiners, and Wiring to BOS Components

After the mounting hardware has been inventoried and sorted out, the next step is to inventory the panels and all the hardware needed to transfer the electricity from the panels down to the BOS components. That also includes all the wiring or cabling used to connect all the components. *NEC Section 690.35* discusses the use of PV wire and PV cable as conductors of PV electricity. USE-2 wire may also be used. Both types of wire must be sunlight-resistant and rated for wet conditions. Most companies making solar equipment now use only PV and USE-2 wire rated for sunlight and wet conditions.

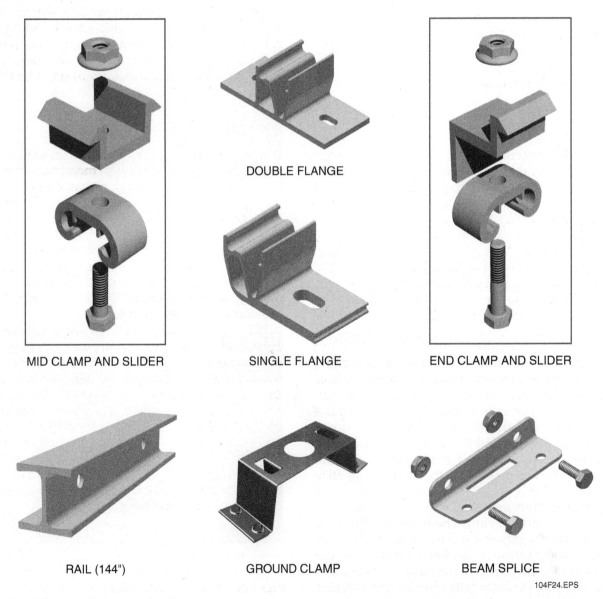

MID CLAMP AND SLIDER   DOUBLE FLANGE   END CLAMP AND SLIDER   SINGLE FLANGE

RAIL (144")   GROUND CLAMP   BEAM SPLICE

104F24.EPS

Figure 24 Solar panel mounting hardware.

4.2.1 Panels and Their Wires or Cables

As shown in an earlier section, the wires from the modules that make up each panel are collected at the junction or terminal box mounted on the backside of each panel. Each of those boxes should have a positive (+) and negative (–) lead coming out of them. Each of those leads (cables) should have a quick disconnect connector on its end. The leads from the terminal boxes are all the same length, usually 6'.

The combiners being used to collect the electrical power from the panels need to be installed close to the panels from which they are collecting. Keeping the combiners close to their panels reduces the voltage drop that can occur in the wiring cables. In some situations, extender cables are needed from the end of the panel wires to the input of the combiner being used. Those extender cables are often bought in pre-cut lengths ranging from 3' to 50'. If one combiner is collecting from six panels, there should be six positive (+) cables and six negative (–) cables with matching quick disconnect connectors on one end. The other end of those cables need to be stripped so that they can be terminated at the input terminals of the combiner. All the cables need to be tagged or labeled with their polarity symbols.

4.2.2 Cable Clamps

All installed cables must be securely fastened to a stationary object to keep the cables from flopping around in the wind. Those stationary objects are usually the mounting beams under the panels. The installation hardware should include a box of cable clamps with the applicable screws. The number of cable clamps cannot be predicted. Local electrical and building codes specify the spacing on the cable clamps.

4.2.3 Combiners and Their Wires or Cables

When the wires from each module inside a panel are tied together in the terminal box on the back of the panel, the output of that box is the total voltage being produced by the panel. When the outputs from several panels are fed into a combiner, the voltage and current both increase. The number of panels being used in an array, and the number fed into a single combiner, determines the number of combiners used on the installation. Each combiner has a positive (+) output and a negative (–) output.

The combiners are normally located on the roof, or wherever the panels are installed. The level of power coming out of a combiner determines the size of power cables used. The output cables from more than one combiner may be fed into another combiner before it sends its output down to the BOS components. The first of the BOS components is a charge controller if batteries are being used. If batteries are not being used, the combiner outputs are fed directly into inverters. There should be a DC disconnect installed between the last combiner output and the first BOS component. If multiple final combiners are used, there may be multiple DC disconnects.

The distance between the last combiner output and the input of the first BOS component will vary, depending upon where each of the components is installed. The wiring or cables from the combiners installed near the panels may or may not be run through metal raceway installed between the last combiner(s) and the first BOS components, depending upon the applicable codes. Newer buildings, those that will have solar arrays installed on them, are being built so that the solar power wiring can be run through the inside of the buildings instead of outside. When PV systems are installed on existing buildings, the wiring metal raceway is usually run outside the building to a point where it can be routed through the roof or the side of the structure and into the inner parts of the building. These installations on older buildings may be called retrofits.

4.2.4 Raceway for Wires or Cables

When the PV system wires are run inside a building, the wires are run through metal raceway. The size and number of wires being run determine the size of metal raceway used. Electrical metallic tubing (EMT) is often used for exterior installations. Metal raceway is normally sold in 10' lengths and is cut to specific lengths as the installation progresses. The number of raceway lengths can be estimated from the plans (such as those shown earlier), but that number will vary a little as the installation is being completed.

Elbow connectors (both 90-degree and 45-degree) are needed also, along with couplings that can connect two sections of raceway. Properly sized clamps and hangers are also needed. Local electrical and building codes specify the spacing on the clamps and hangers. The elbow connectors, couplings, clamps, and hangers may be bought individually, but are often bought in bulk for large installation projects. If the drawings and plans are specific enough, the exact number of connectors, clamps, and hangers may be determined, but do not expect such detailed plans.

4.3.0 BOS Components

If the general arrangement and electrical drawings shown earlier are reviewed again, the exact location of each component used for that project is shown. The EX-2 drawing shows two combiner boxes. Their physical size is noted on the drawing, along with their NEMA rating. Each combiner is located near its respective panels (called modules on these drawings). This drawing also shows the distance from the combiners to the PV power source disconnect (DC disconnect) located near the inverter. Notice that all the disconnects and the BOS equipment are located outside the structure. In this example, all of the BOS equipment is within 10' of the main service panel. Note 4 on the EX-1 electrical drawing gives the NEMA rating and capacity of the DC disconnect switch.

As for the BOS components themselves, their details are shown on the EX-1 drawing. Notes 5, 6, and 7 give the details about the Sunny Boy inverter being used. This particular inverter has ground fault protection built into it. It is positioned within 10' of the main service panel. On the output of the inverter is a PV output meter. Beyond it is an inverter output disconnect switch. This switch is the AC disconnect switch that removes the PV power from the main service panel. According to Note 8 on the EX-1 drawing, the inverter output disconnect does not have to be installed if the utility disconnect is within view of the inverter. After being checked during the inventory, all the BOS components should be stored in their shipping containers until they are ready for installation. Keep all the manufacturer's instructions that come with each BOS component. This information is needed later when the components are installed.

5.0.0 INSTALLING MOUNTING HARDWARE AND RACEWAYS

PV panels can be installed almost anywhere. Spacecraft have been using solar power for years. In the last few years, people have built and flown airplanes whose sole power comes from solar panels installed in their wings. PV systems are being used more and more on pleasure boats and recreational vehicles. Companies like Wal-Mart with large flat-roofed buildings are beginning to have PV arrays installed on top of those large open spaces. Some utility companies are leasing the roofs of major warehouses and having solar farms installed on top of them. Regardless of where the solar panels are being installed, they all will be mounted in one of these methods: integral mounting, standoff mounting, ground mounting, or pole mounting. Some mounting hardware is movable (manually or by drive motors) so that the arrays can be repositioned to follow the sun's movements during the day or during the seasonal changes. *Figure 25* shows samples of each of the four mounting methods most often used.

INTEGRALLY-MOUNTED PANELS

GROUND-MOUNTED PANELS AS CARPORT COVERS

STANDOFF-MOUNTED PANELS

POLE-MOUNTED PANELS

104F25.EPS

Figure 25   Different mounting methods.

NCCER — *Contren® Learning Series* 57104-11

Integrally-mounted panels are built into the roof or exterior wall of a building. They are sometimes called building-integrated photovoltaics, or BIPVs. Standoff mounting uses legs or standoffs to support the mounting beams or rails on which the panels are mounted. The standoff mounts can position the panels parallel to the roofing surface, or they can be cut to different lengths so that they will tilt the panels toward the sun. Regardless of their lengths, the standoffs must be anchored. In most cases, they are anchored to the structural members supporting the roof, or whatever surface on which they are mounted. In other situations, the frames or racks on which the panels are attached may be just sitting on a surface with large ballast (weights) placed on their bottom rails. Regardless of how they are anchored, they normally keep the panels at least a few inches above the mounting surface, but some are raised more than that. Solar panels, other than the integrally-mounted ones, need to have air movement around their undersides. The moving air helps keep the panels cooled. When array panels are tilted, one edge is high and the other low. The different rows of panels must be spaced apart so that they do not shade one another.

Ground-mounted arrays are similar to those mounted on a building roof or its side, but they must be treated differently. Some kind of fencing needs to surround them to keep out intruders and vandals. Fences also keep animals or unqualified personnel from getting shocked by any of the equipment. Ground-mounted arrays are attached to structural members called racks, or attached to racks mounted on a single post.

Most ground-mounted panels are installed on frames or racks. The racks tilt the panels toward the sun. The racks are usually bolted or welded to metal posts sunk into concrete pillars buried in the ground. Some ground-mounted racks are held down by ballasts instead of ground anchors (*Figure 26*). Pole-mounted arrays also have racks attached to metal poles sunk in concrete pillars, but the pole-mounted arrays only have a single post. While most ground-mounted arrays are stationary, some pole-mounted arrays can be moved in elevation (tilted up or down) and azimuth (moved left or right). Such movements allow them to be more focused on the position of the sun as the seasons change. Most adjustable arrays are moved manually, but some have electric motors on them that move them in azimuth and elevation. Motorized arrays are much more expensive.

This section focuses on preparing and installing the following items on a house with a shingled roof:

- Mounting hardware for four solar panels mounted at least 4" above roof

BALLAST

Figure 26 Ballasted array.

104F26.EPS

- Four solar panels
- One combiner with wiring from the four solar panels
- Metal raceway and wiring from the combiner to a DC disconnect switch
- One charge controller and at least two FLA L-16 batteries
- Metal raceway from the DC disconnect switch to the charge controller
- Metal raceway and wiring from the charge controller to the batteries
- One inverter
- Metal raceway and wiring from the batteries to the charge controller and/or to the inverter
- One AC breaker inside an enclosure
- One AC load (a light fixture)
- Metal raceway and wiring from the inverter to the AC breaker
- Metal raceway and wiring from the AC breaker to the AC load (light fixture)

Figure 27 shows a simple diagram of the items to be installed for this system.

NOTE

The panels can be wired in parallel or in series. For this exercise, they will be wired in series and in parallel. Some codes require a DC disconnect near the panels and one near the ground floor. An optional DC disconnect can be added just prior to the charge controller.

5.1.0 Preparing Installation Site

Follow OSHA safe work practices and company safety policies when preparing the site for installation of a PV system. Always perform a job safety analysis (JSA) before proceeding with any work.

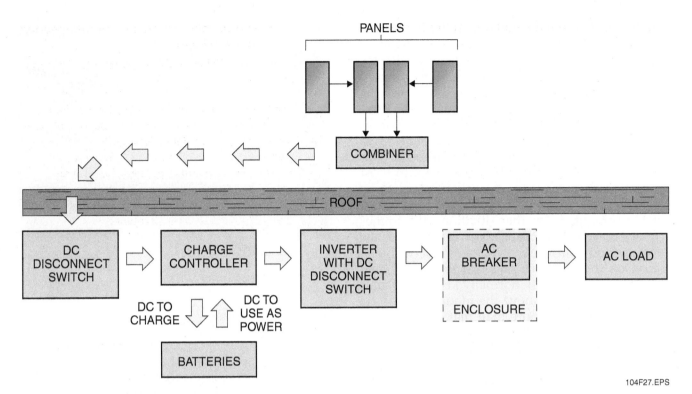

PANELS

COMBINER

ROOF

DC DISCONNECT SWITCH

CHARGE CONTROLLER

INVERTER WITH DC DISCONNECT SWITCH

AC BREAKER

ENCLOSURE

AC LOAD

DC TO CHARGE

DC TO USE AS POWER

BATTERIES

104F27.EPS

Figure 27   Simple system diagram.

A JSA examines all of the hazards associated with a task and provides specific steps for performing the work safely. A sample JSA form can be found in the *Appendix*.

The following are steps to take in preparing a site for the actual installation:

Step 1   Verify that all personnel are wearing the correct PPE and fall protection equipment for roof work.

Step 2   Verify that any scaffolding or ladders being used are properly placed.

> **NOTE**
>
> If an aerial lift must be used, verify that its operator is properly trained and certified.

Step 3   Inspect the lifting device that will be used to transport the materials to the roof. Verify that it is safe to use. Test it without a load to make sure it works.

Step 4   Locate or install all necessary anchor points that will be used for lanyard tie-off points on the roof. Verify that they meet minimum requirements.

Step 5   Review the plans, drawings, and permits one more time to make sure where the panels must be mounted. Make sure there is enough room around the perimeter of the proposed location to meet all building and safety codes.

Step 6   Locate all rafters (or trusses) that will be used to support and anchor the panels. Make sure that they do not need any additional bracing. Some installers use a heavy rubber or leather mallet to tap the roof to find where the rafters are located. Others use a strong stud finder to find the nails in the rafters.

Step 7   Identify the rafter(s) closest to the outer edges of the installation area, and mark a high point (near the ridge) and low point (near gutter) on it (or them).

> **NOTE**
>
> Remember that reference holes can be drilled up through the rafter from the inside – if there is enough room in the attic. Working from the marked rafter, an installer can locate and mark all the other rafters.

Step 8   Verify the centering distance between the rafters or trusses that will be under the installed panels. Transfer those measurements onto the roof and mark the high and low points on each rafter.

Step 9 Snap a chalkline on each rafter to mark its location.

Step 10 Review the plans and drawings again to determine the following about the panels and their mounting hardware:

- How the panels will be mounted (portrait or landscape).
- How the panels will be connected (in series or in parallel).
- How far the panels will be in from the ridge and gutter areas of the roof.
- Where the outer edges of the panels will be.

- The number of anchor points needed for each beam.

Figure 28 shows the layout of the panels.

> **NOTE**
>
> For this example, the beams will be installed perpendicular to the rafters. Each beam will have three anchor points (standoffs). The panels will be installed in series, but in a portrait position and in parallel. Two panels will be connected in series. Two more will be connected in series. Those two series will be tied, in parallel, into the combiner.

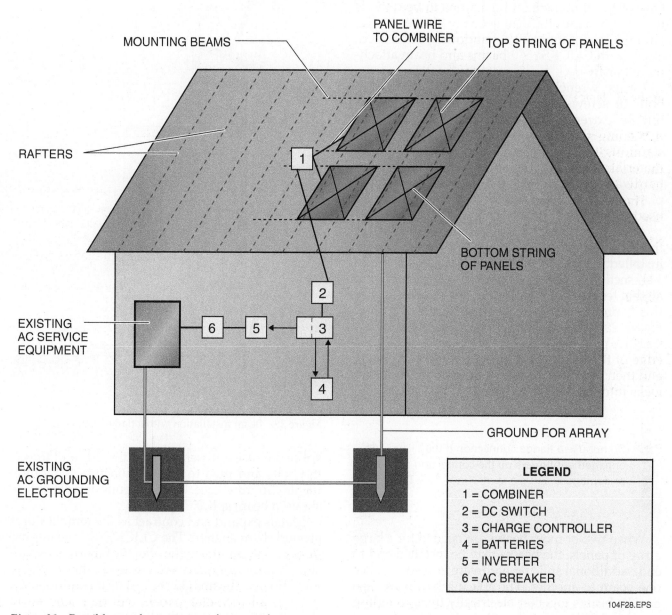

104F28.EPS

Figure 28 Possible panel arrangement for example system.

 System Installation and Inspection

Step 11 Locate and mark the rafters that will be used to anchor the standoffs. Mark exactly where the location of the holes need to be drilled for the standoff lag bolts or screws. When marking rafters, remember that the edges are actually less than 2" thick. The anchors must be in the middle third of the rafter.

CAUTION

Do not drill unnecessary holes in the roof. Remember that any hole drilled into a roof must be properly sealed when the work is completed.

5.2.0 Installing Mounting Hardware

One of the first installation actions to be done is to get the standoffs installed. For this example, Unirac Quick Mount PV® standoffs will be used, along with CLICKSYS™ beams and beam attachment hardware.

Torque wrenches are used to tighten the lag bolts or screws to specified amounts of torque, but be extremely careful when installing and tightening the lag bolts or screws that tie the standoffs to the rafters used under the roofing materials. Refer to the vendor manual for the hardware being installed. The site assessment plan may have indicated the type of wood being used for rafters.

After the standoffs are installed, the beams that will actually be holding the panels can be installed. Unirac or CLICKSYS™ beams come in 144" sections. Some installers may shorten them if all that length is not needed for an installation. For this training, complete beams will be installed. *Figure 29* shows the three steps used to lock a CLICKSYS™ beam into a flange connection. One edge of the beam is first inserted into the flange, and then the other is pressed down until the beam locks into the flange connection.

CAUTION

Do not use a flange connection if the engagement fingers on the connection are bent or damaged.

When two or more beams are needed for a large array of panels, the beams may be installed end to end. Additional splicing hardware is used to splice one beam to another. The splicing hardware also ties the beams together electrically (for grounding purposes). When installing the splicing bracket and

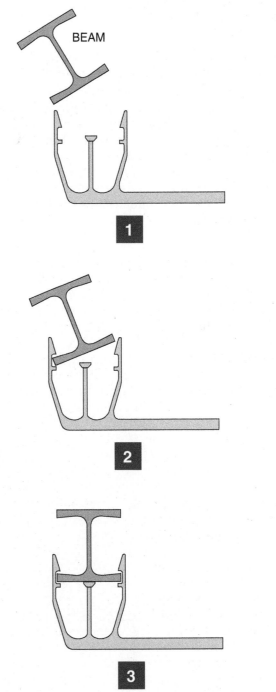

104F29.EPS

Figure 29 Beam installation with a flange connection.

splice bar, use anti-seize on the threads and torque the bolts and nuts to the manufacturer's requirements (in this case, 10 foot-pounds). *Figure 30* shows a beam splice.

Metals expand and contract as the temperatures around them change. The CLICKSYS™ beams are designed to minimize the effects of thermal expansion, but the company recommends that no more than three full beams (36') be installed continuously.

Since all the solar panels will be anchored to the mounting beams, the beams must be tied to

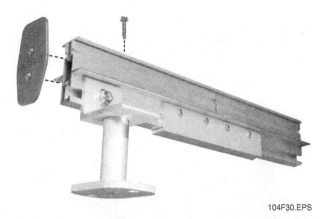

Figure 30 Beam splice.

104F30.EPS

104F31.EPS

Figure 31 Grounding lug.

a ground for grounding purposes, fault protection, and lightning protection. *Figure 31* shows a grounding lug attached to one of the beams.

After the beams are securely installed and grounded, the hardware that will actually hold the panels in place can be installed. Clip-like devices called sliders are slid over the ends of the beams and moved into the positions needed for installing the panels. Sliders have threaded bolts. End clamp brackets installed over the slider bolts anchor the panels to the beams. When two panels are installed side by side, a clamp called a mid clamp is used. *Figure 32* shows the installation of the sliders, end clamps, and mid clamps.

For this example, a total of six beams will be used. Three of the beams will be installed parallel to each other for one set of panels. Three more beams will be mounted (parallel to each other) and parallel to the first set of beams. The spacing of the parallel beams will depend upon the amount of loading expected on the panels. The third beam adds additional support if the system designer is expecting high wind loading. Each beam will have three Quick Mount PV® standoff assemblies (flashing, standoff blocks, and lag bolt/screw with a type 1 flange connection). *Figure 33* shows the layout of beams, panels, and clamps.

Before installing the mounting hardware, develop a JSA that will be followed for this task. Perform the following steps to install the mounting hardware for four solar panels positioned as noted in *Figure 33*:

Step 1 Obtain the following materials and tools:
- 4 beams
- 3 sets of UNIRAC Quick Mount PV® standoffs (or suitable substitutes) per beam (total of 12 sets)
- At least 16 sets of end clamps, and at least 6 sets of mid clamps
- Measuring tape, markers, framing square, level, stud finder, and chalkline

- Drills (battery-powered drills preferred over corded drills) for drilling and bolt or screw insertion or removal
- A few long ¼" drill bits made for drilling wood and shingles
- Deep-well, thin-wall, sockets to fit the lag bolts or screws and the nuts used to secure the panel clamps (the sockets also need to fit the drive of the torque wrench)
- Roofers flat pry bar for lifting shingles to insert flashing
- Caulking gun with roofing sealant that is compatible to the roofing material
- Torque wrench capable of measuring up to 200 inch-pounds
- Some form of marker or marking material to use on bolts that have been torqued

WARNING!
Remember that all tools and materials must be kept under control at all times on the roof. Use secured containers to hold trash. If possible, some kind of fine-mesh netting should be erected below the work area to catch any tools or materials that are dropped.

Step 2 Transport the mounting materials and tools to the roof, and secure them.

WARNING!
Do not climb ladders with materials or tools. Use the material lift device. If no lift is available, tools and small parts can be placed into a bucket that can be lifted via rope to the roof. The mounting materials also can be pulled onto the roof via a rope.

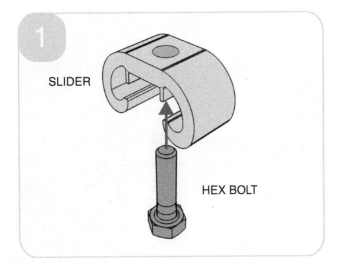

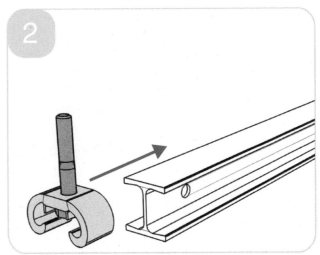

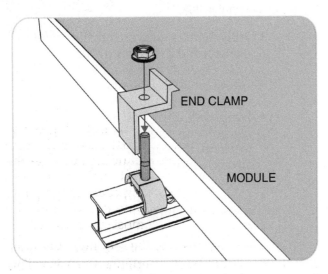

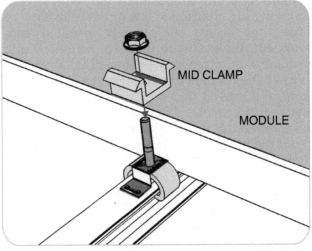

104F32.EPS

Figure 32   Installation of sliders, end clamps, and mid clamps.

Step 3   While facing the sloping roof and the ridgeline, locate the point where the first standoff is to be installed.

Step 4   Use one of the beams as a straightedge and locate the second and third standoff installation points for the beam. Verify that the beam will be level if the standoffs are properly installed.

Step 5   Position the drill bit so that it is perpendicular to the surface of the roof and drill the first pilot hole.

Step 6   Use the flat pry bar to carefully lift the shingle(s) over the drilled hole, and then squeeze a dab (small amount) of caulking into the drilled hole.

Step 7   Slide the flashing under the shingles, align its bolt hole with the drilled hole, and remove the pry bar.

Step 8   Insert the lag bolt/screw through the flange connection (*Figure 34*) and then slip the tip of the lag bolt/screw through the standoff block into the drilled pilot hole. Tighten the lag bolt/screw until most of the bolt/screw is in the rafter, but the flange connection is still movable. Do not tighten yet.

Step 9   Drill a pilot hole at the next standoff location and repeat the process until all standoffs have been installed for the first beam.

Step 10   Position the beam above the three flange connections and rotate the beam so that one of its bottom edges drops down into the locking slot of the flange. The loose flange connections may need to be slightly repositioned to get all the fingers aligned to catch the beam. Press down on the beam and snap it into place.

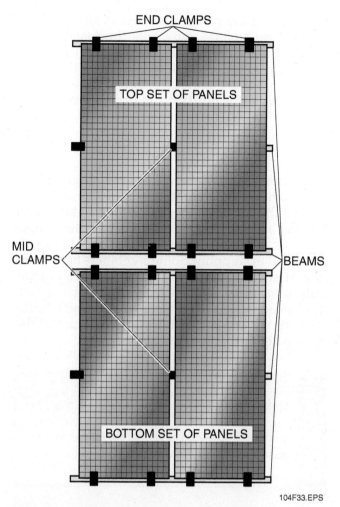

Figure 33   Layout of twin sets of solar panels.

104F33.EPS

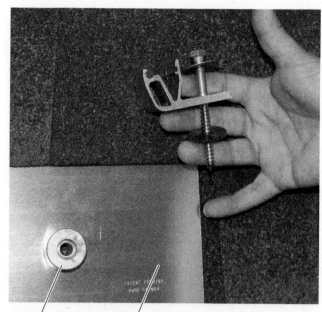

STANDOFF   INSTALLED FLASHING

104F34.EPS

Figure 34   Flange and flashing.

Step 16 Determine how many feet of grounding cable will be needed to reach from the array to the grounding stake on the ground, plus an additional 10' of cable. Attach the grounding cable to the panel beams, and then route the cable off the roof by way of the most direct route.

Step 17 Mark the threads of the installed bolts and nuts using a permanent marker. This will show if they loosen over time.

> **NOTE**
>
> At this point, the panels can be installed. If the wires from the panels need to be run into or through metal raceway that will be installed under the panels, this would be the time to install it. If the panel wires will be extended out from under the panels and connected to the combiner and a disconnect switch, the combiner and switch need to be installed at this time. Metal raceway from the disconnect switch can also be run from the roof down to the BOS components at the ground level.

Step 11 Tighten the lag bolts/screws to the desired torque level.

Step 12 Install the second and third beams in the same manner.

> **NOTE**
>
> At this point, the beams for the first set of panels have been installed. The beams for the next set of panels will be installed in the same manner. After the beams are installed, the sliders can be slid onto the beams.

Step 13 Install the three beams for the second set of panels.

Step 14 Assemble the sliders for each beam and slide them into their approximate position.

Step 15 Attach the ground lugs to the beams.

6.0.0 INSPECTIONS OF PV COMPONENTS PRIOR TO INSTALLATION

Any time new equipment is received, it is best to first visually inspect it for damage before installing it or performing any operational checks on it. This is especially true for electronic components. The electronic components being built today are much sturdier than those of years past, but they still can be damaged if dropped or otherwise mishandled. Static discharges (from humans, packing materials, other machines, or welding activities) can damage the electrical circuitry inside the components. They can also be damaged by exposure to high levels of electromagnetic forces. The following sections start with the PV panels and go all the way down to the AC circuit breaker that will be used to transfer the PV electrical power into the main service to the building.

6.1.0 Inspecting PV Panels

As noted earlier, solar panels create direct current voltage when they are exposed to light. When delivered to an installation site, the panels are normally covered and boxed. They need to stay covered until they are to be installed. If an installer needs to verify that a solar panel is operational, a volt meter can be used across the positive (+) and negative (–) leads coming from the terminal box mounted on the back side of the panel. *Figure 35* shows a close-up view of the nameplate data on the back of a Canadian Solar PV panel. Notice the positive (+) and negative (–) markings on the wires.

The voltage read across the panel will be less than the level of voltage shown on the nameplate, especially if the panel is shaded at all. Checking a panel for voltage only tells the installer that the electrical circuits (cells and wiring) inside the panel are functional. Make sure that the wires from a panel are undamaged, and that they are marked with the positive and negative tags.

6.2.0 Inspecting Combiner and DC Disconnect Switch

The output wires of the solar panels should never be attached to the combiner until the combiner and all the circuits between it and the AC circuit breaker are wired in. Most combiners are simple terminal boxes or enclosures that should not be damaged in shipping. Some combiners have fuses mounted between their input terminals and their output terminal. If a combiner does have a fuse across each input, then a continuity check could be made (before any wiring is attached) across the fuse to verify that the fuse is functional.

> **WARNING!**
>
> Make sure that there are no inputs into the combiner before making any continuity checks on the combiner fuses or the disconnect switch downstream from the combiner.

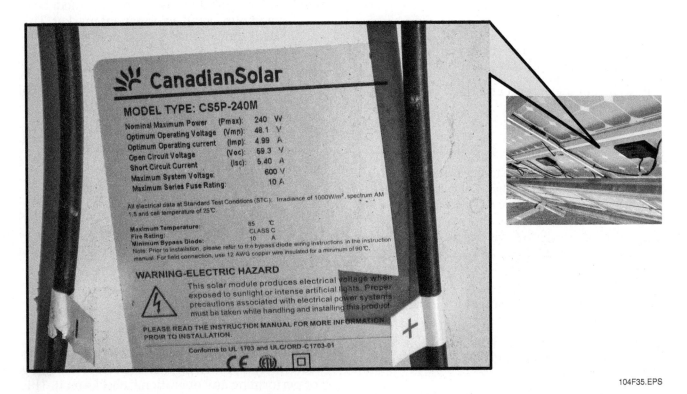

Figure 35 Nameplate data on back of a PV panel.

104F35.EPS

The output of the combiner is usually fed into a DC disconnect switch that will normally be mounted somewhere near the combiner. Some combiners may have a disconnect switch built into them. The disconnect switches are designed to stop the output of the combiner from going on down the line to the other PV circuits. Open and close the contacts to verify proper operation of the disconnect switch.

6.3.0 Inspecting Charge Controller and Batteries

The charge controller is a much more complicated device. Each charge controller is a little bit different, depending upon its maker. How they operate depends upon how they are used in a PV system. If the PV system is an off-the-grid, stand-alone system, the charge controller has one set of tasks. If the PV system is tied into the grid, then the charge controller has a different set of tasks. Review the site design plans and drawings to determine the type of charge controller that will be used, and then review the manuals that come with the charge controller to see what checks the maker recommends. Until a charge controller is actually installed, there is not much that can be done other than visually verifying that it has not been damaged in shipping. It may require specific programming when installed. Make sure to keep the charge controller away from any strong electromagnetic forces or contacts with any kind of electrostatic discharges that could damage its electronic control circuits. In some situations, anyone touching anything inside the charge controller should be wearing electrostatic discharge (ESD) grounding straps on their hands or arms. Some situations also call for the person to be standing on an ESD mat. *Figure 36* shows a technician wearing an ESD strap while working inside a control cabinet.

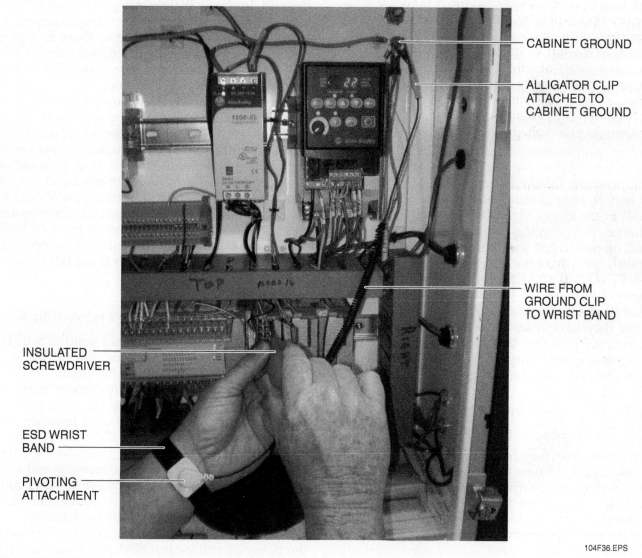

104F36.EPS

Figure 36 ESD strap.

Batteries are being addressed here with the charge controllers because the two work together. Sealed batteries too are pretty much plug and play devices, but FLA batteries are a different story. FLA batteries, such as the often used L-16 batteries, are normally shipped to the site dry. At some point, an approved battery-grade electrolyte (acid mixture) must be added to the battery cells to activate them. If that is the case, a hydrometer is needed to check the specific gravity of the electrolyte in the battery. After the electrolyte has been added to the battery cells, the batteries must be allowed to sit for a time specified by the battery maker. Most FLA batteries need 90 minutes to two hours for the plates and other components inside the battery to absorb the electrolyte. During this activation time, the electrolyte's temperature increases and its specific gravity decreases. After the activation period has ended, the batteries can be placed onto a charger at the finishing rate recommended by the battery maker.

The temperature of the battery cells needs to be monitored during the charging process. The temperature should never exceed 115°F (46°C). If the cell temperature becomes excessive, or the cells begin to gas vigorously, reduce the rate of charge. The charging process continues until the cell or cells come to within 0.005 points of the specific gravity of the filling electrolyte, corrected for 77°F (25°C). The Surrette-Rose Battery Company, one maker of L-16 batteries, recommends that the batteries be charged an additional hour after they have reached their initial charge. The additional charging time ensures no further rise in specific gravity. After the batteries are prepared and charged, they can be placed into service.

6.4.0 Inspecting Inverters and AC Circuit Breakers

Inverters, like the charge controllers, are more complicated because they contain electronic control devices. Inside inverters, these electronic control devices convert DC power into AC power. When inverters are shipped to the site, they should still be in their protective shipping boxes. The boxes should have been stored until the inverters are about to be installed. If an inverter will be installed soon, it should be taken out of its shipping and storage packaging and inspected. There should be no apparent damage to the inverter housing. Attached cables and wires should be securely attached with no apparent damage. Those cables and wires that must be plugged into the inverter housing should also appear to be undamaged. Make sure to keep the inverter away from any strong electromagnetic forces or contacts with any kind of electrostatic discharges that could damage its electronic control circuits. Check the vendor manual on the inverter to see what ESD equipment is required, if any, when working on and even checking the control circuits in an inverter.

Some inverters may have charge controllers built into them so that they can service any batteries, as well as converting the DC to AC voltages. To comply with the *NEC*®, all inverters must be protected by ground fault circuit interrupters (GFCIs). If a ground fault happens in the DC wiring, the inverter is supposed to disconnect from the grid. Inverters that tie into grid-connected circuits are designed to monitor the grid power. If grid power becomes unstable or there is a grid outage, the inverter will shut down. Review the inverter manual to see if there are any checks that can be made prior to installation.

The output of an inverter should be an AC voltage representing the power being produced by the PV system. When that output power is fed into an AC circuit breaker, that breaker simply operates as a normal circuit breaker. If an ohm meter is placed across the AC breaker input and output contacts prior to the breaker being installed, the meter should show a short when the contacts are closed. The meter should show an open or high resistance when the breaker contacts are opened. Make sure that the AC breaker is the last component installed.

6.5.0 Inspecting Raceway and Wiring

This module assumes that rigid conduit will be used as the raceway from the combiner all the way to the enclosure to house the AC circuit breaker. Installers checking the PV equipment need to review the site design plans to see what kind of raceway and wiring must be used between the PV components. The installers also need to review all the electrical and building codes to see what applies to the installation of the raceway and wiring. Tools will be needed to measure, cut, bend, install, and secure the raceway. Remember that the larger conductors require larger raceway.

Determine where the charge controller and the batteries will be installed and then measure the distance from the combiner to the charge controller. Make notes as to the number and types of bends that will be needed between the two. Also determine the type and lengths of wire that will be needed. Sketch out the run. Determine the type and number of clamps or hangers needed to secure the raceway. If the raceway will have to penetrate the roof, a properly-sized drill bit will be needed. Roofing sealant will also be needed if a hole must be drilled or cut in the roof.

Determine where the inverter will be installed, and then determine the amount of raceway and wire needed to connect the charge controller and the inverter. Do the same for the raceway and wire that will be run from the charge controller to the batteries. The enclosure that will house the AC circuit breaker may be installed some distance away from the inverter. Determine the amount of raceway and wire needed between those two items.

After all the measurements have been made, and the calculations have been made for the needed bends in the raceway, check the supply of raceway delivered to the job. Make sure that there is enough raceway to do all the installations. Also, check the supplied wire.

7.0.0 Installing PV Modules and Balance-of-System Components

Now that all the PV system components have been checked, and the mounting beams for the solar panels are ready, the actual installation process can begin. Since the solar panels can be easily damaged, and they do not need to be wired into the PV system components until the last thing, they will be the last items installed. Review the site assessment report and the system design plans to verify where all the PV components are to be installed. The following is the sequence of installations:

- Installing the combiner
- Installing a DC disconnect
- Installing the charge controller
- Installing the batteries
- Installing the inverter
- Installing the PV panels

7.1.0 Installing the Combiner

Before attempting to install the combiner selected for the PV system, take time to review the user's guide that came with the combiner. It gives specifics about the combiner, where it can be installed, general instructions, cautions, warnings, and instructions on how to wire the device. For this exercise, assume that the combiner will be installed outside on a roof that slopes 20 degrees.

Some combiners must be mounted vertically to keep them rainproof while others can be leaned back a few degrees. The following are general installation steps based on an Outback Power System PV 8 combiner. Other combiners may have more or less inputs, but they all are very similar.

This Outback Power System PV 8 combiner can be installed vertically or leaned back 14 degrees. It has multiple knockouts for both mounting and connecting the wiring. The terminals and breakers or fuse holders are installed after the enclosure has been mounted. The inputs enter the bottom of the enclosure and the outputs exit the top.

Figure 37 shows the general location of components, inputs, and outputs of a combiner equipped with fuses. Another version of this combiner uses circuit breakers instead of fuses. Each input from a string of solar panels passes through one fuse or circuit breaker before combining with the other inputs on a combining bus at the top of the combiner.

In most cases, the DC+ and DC– wires from the individual solar panels are not long enough to reach the combiner. Extender wires are usually used between the wires from the panels and the combiner. The extender wires need to have matching quick-disconnect connectors on the ends mating with the panel wires. The other end of the extender wires will be stripped wires that can be attached to the terminal screws (lugs) inside the combiner.

> **WARNING!**
>
> To prevent personal injury caused by electrical shock, make sure that the wires from the panels are not connected to any other wiring.

The combiner generates a single DC+ output and a single DC– output. Those outputs are the two large black cables extending from the top of the circuit breakers down through the combiner. They exit at the bottom of the enclosure. The sheet metal that makes up the enclosure extends below the input and output connections. It provides weather protection to the connections.

Before installing the combiner, develop a JSA for this task. Follow these steps to install a PV system combiner:

Step 1   Determine the number of input and output connectors needed in the combiner enclosure.

Step 2   Mount, level, and secure the combiner.

 System Installation and Inspection

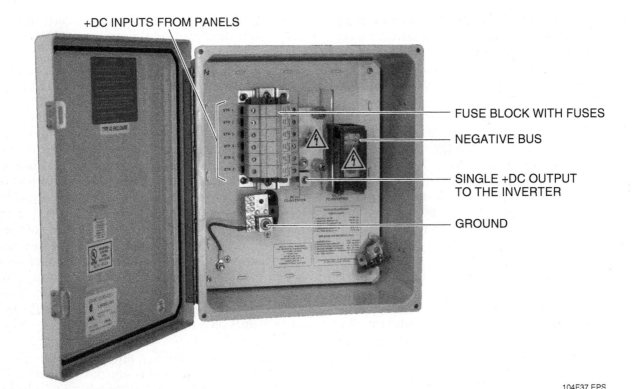

+DC INPUTS FROM PANELS

FUSE BLOCK WITH FUSES

NEGATIVE BUS

SINGLE +DC OUTPUT
TO THE INVERTER

GROUND

104F37.EPS

Figure 37   Components, inputs, and outputs of a typical combiner.

Step 3   Slide the circuit breakers or fuse holders onto the DIN rail inside the enclosure and lock them in place.

Step 4   Install and secure the box lugs to the combiner busbar.

Step 5   Install the combiner busbar onto the top of the breaker assembly (or the fuse holder assembly).

Step 6   Install protective bushings for the array wires entering the enclosure through the bottom knockouts.

WARNING!

To prevent electrical shock, make sure that the extender wires from the panels are not attached to the panel outputs at this time.

NOTE

Raceway is not normally used for the wires coming into the combiner from the panels, but the wires have to be properly secured to keep them from moving around after installation. Ensure that all wires not inside raceway are secured in accordance with all codes.

Step 7   Route the array wires through the protective bushings and into the combiner. Pull in enough array wire to reach the terminals at the bottom of the circuit breaker assembly or the fuse holder assembly (whichever is used).

Step 8   Cut the incoming wires to a length that will allow them to be easily terminated on their respective terminals.

Step 9   Attach the negative wire to one of the lugs on the negative terminal busbar and the positive wire to the bottom lug of a circuit breaker (or fuse holder).

Step 10   Torque each lug as specified by the equipment manufacturer.

CAUTION

To ensure a good electrical connection, make sure that all lugs are tightened to the specified torque level. To prevent damage to the equipment, do not overtighten any lugs.

Step 11   Obtain enough properly sized ground wire to tie the combiner to the system ground. Review the system design plans for specific grounding information.

Step 12 Install and secure the ground wire to the combiner's ground busbar and the overall system ground.

Step 13 Determine where the DC disconnect switch will be mounted, then obtain the metal raceway and associated connectors and elbows needed to connect the combiner to the DC disconnect switch.

Step 14 Install the metal raceway from the combiner to the DC disconnect switch.

Step 15 Obtain enough wire to reach from the combiner terminals to the DC disconnect switch terminals (twice), plus an extra few feet of wire. Remember that there will be a DC+ wire and a DC– wire. Also, obtain enough grounding wire to tie the two together.

Step 16 Pull the wires through the metal raceway between the combiner and the DC disconnect switch.

Step 17 Terminate the DC+ and DC– wires at the combiner. Tie in the ground wire.

> **WARNING!**
>
> Before the panels are connected to the combiner and DC power is allowed to flow, all wiring must be rechecked by a licensed electrician.

7.2.0 Installing a DC Disconnect Switch

The *NEC®* requires that PV systems have a means of disconnecting the inverter(s) from the DC power supplied by the panels. The combiner covered earlier may have been equipped with either a circuit breaker or a fuse on each input into the combiner. Some combiners may not be so equipped. While one of the circuit breakers could be used to stop the power coming from a single string of solar panels, this is not a recommended process. The recommended way to meet the code requirements is to insert a dedicated safety disconnect switch designed to handle the DC power of a PV array (*Figure 38*).

These Square D Class 3110 safety switches have multiple uses in industry, but they must be wired a specific way for photovoltaic use. The *NEC®* does not allow the negative conductor to be switched when disconnecting PV systems. *Figure 39* shows how this Square D disconnect must be wired for a PV system.

104F38.EPS

Figure 38 DC disconnect switch.

According to the Square D bulletin, one inverter may be connected to each output pole of this Class 3100 disconnect switch. The bulletin goes on to say that if these safety switches are not fused, the inverter (downstream from the switch) must not be capable of backfeeding currents into a short circuit or fault in the PV array or string. What that means is that these switches are direct connections when they have no fuses. A short downstream could backfeed through the contacts of this switch and damage the panels. *Table 3* shows the electrical limitations (current ratings) for these Square D disconnect switches.

These DC disconnect switches are normally mounted outside and near the combiners. They must be rainproof.

Before installing the DC disconnect switch, develop a JSA for this task. Follow these steps to install a DC disconnect switch:

Step 1 Prepare the DC disconnect switch enclosure so that the metal raceway from the combiner can be installed. Install the metal raceway connectors.

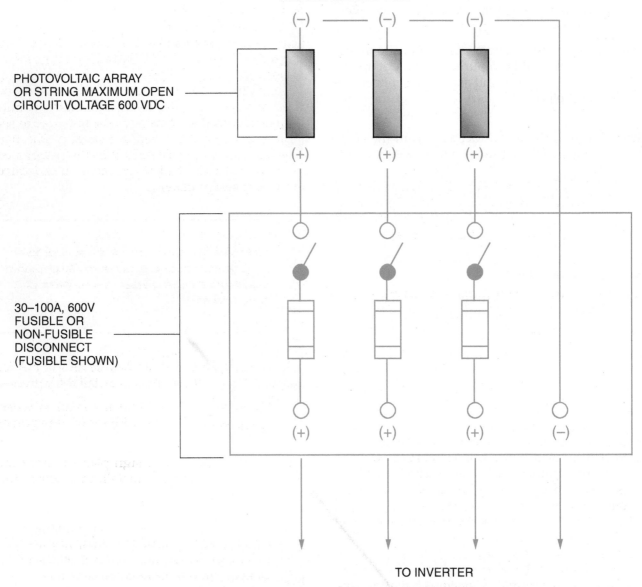

PHOTOVOLTAIC ARRAY
OR STRING MAXIMUM OPEN
CIRCUIT VOLTAGE 600 VDC

(−)   (−)   (−)

(+)   (+)   (+)

30–100A, 600V
FUSIBLE OR
NON-FUSIBLE
DISCONNECT
(FUSIBLE SHOWN)

(+)   (+)   (+)   (−)

TO INVERTER

- Connect negative photovoltaic line (−) to grounded inverter.
- Positive grounded systems are similarly allowed.
- For ungrounded systems, see *NEC Section 690.35*.

104F39.EPS

Figure 39 DC switch wiring for PV systems.

Table 3 Current Ratings

Switch Nameplate 600V	Switch DC Rating per Pole[1]	Photovoltaic Maximum Circuit Current[2]	Photovoltaic Short-Circuit Current (I_{sc})
30A	20A	16A DC per pole	12.8 (20/1.56)
60A	60A	48A DC per pole	38A (60/1.56)
100A	100A	80A DC per pole	64A (100/1.56)

Step 2 Feed the wires into the DC switch enclosure and the secure the metal raceway from the combiner to the enclosure.

Step 3 Level and anchor the DC disconnect switch at its mounting location.

Step 4 Prepare the ends of the combiner wires for termination at the DC disconnect switch.

Step 5 Determine if the DC+ input will be split and run through multiple DC+ terminals of the disconnect switch.

Step 6 Terminate the DC+ and DC– wires coming into the DC disconnect switch. Tighten the wire lugs to level of torque specified by the maker of the disconnect switch.

NOTE
The metal raceway from the switch to the charge controller will be installed later. The wires will be pulled after the charge controller is ready.

7.3.0 Installing the Batteries

Before installing the charge controller(s) and the inverter(s), a decision must be made about where to install the batteries for the PV system. Review the site design plan to see where to install the batteries. That plan also shows how the batteries will be wired. *Figure 40* shows a drawing with four 6V batteries tied in series to a charge controller. The result is a battery bank that will give up 24 VDC when fully charged. Notice that the negative (–) terminal of the top battery is tied through a shunt to a DC negative busbar. Notice also that B– terminal of the charge controller is also tied to the DC negative busbar. The controller's B– terminal relates to the negative side of the batteries. The positive (+) terminal of the bottom battery is tied through a circuit breaker to the BAT+ terminal on the charge controller.

The batteries must be installed in a cool area that is well ventilated. The cables used between the batteries and the charge controller need to be kept as short as possible. To reduce the risk of arcing as the cables are being connected between the battery bank and the charge controller, the battery cables need to be secured to the batteries before they are attached to the charge controller. Always keep the bare ends of the cables covered until they are stripped to be attached at the controller. Some codes may call for the battery cables to be run through metal raceway to the charge controller. Check local and national codes for any

further guidance on metal raceway used between batteries and their charge controller(s).

As noted earlier, the batteries will be either FLA batteries or some kind of sealed battery. This project will use two FLA batteries tied in series to produce 12 VDC. These FLA batteries are normally shipped dry and activated by the addition of electrolyte at the site. Refer to the user manual or instructions that come with the batteries to see how to activate them. Upon receipt of the batteries, install them as directed in the designated location. Hook up the battery cables as indicated in the site design plans.

WARNING!
Make sure to comply with all safety guidelines for working with or near batteries. Do the same if acids must be handled or poured. Wear all appropriate PPE.

Before installing the batteries, develop a JSA for this task. Follow these steps to install the batteries:

Step 1 Ensure that everyone involved is wearing the appropriate PPE for working on or near batteries.

Step 2 Review the site design plan to verify the type of batteries to install, and their installation location.

NOTE
If the batteries are sealed batteries, they can be installed without much other effort. If the batteries are FLA batteries that need to be activated, follow the manufacturer's instructions on how to activate them. After their activation process has been started, they can be installed and hooked together.

Step 3 Verify that the battery installation site has good ventilation, and that the batteries will be kept out of the sunlight.

Step 4 Place the batteries in their installation location and make sure that they are secure.

Step 5 Review the site design plan to verify how the batteries should be wired (together). They will be either wired in series or in parallel. For this exercise, they will be wired in series.

Step 6 Install the cables that will be used to tie the batteries together.

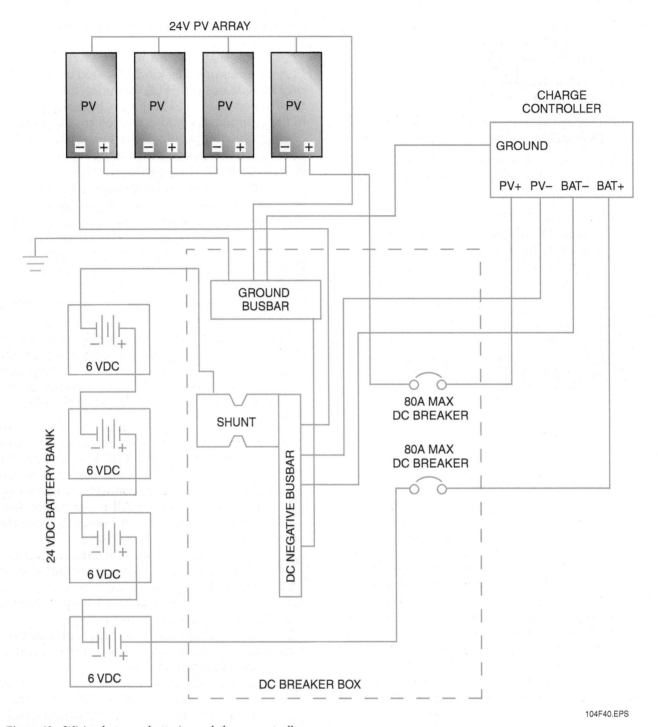

24V PV ARRAY

PV PV PV PV

CHARGE CONTROLLER

GROUND

PV+ PV− BAT− BAT+

GROUND BUSBAR

6 VDC

6 VDC

SHUNT

6 VDC

6 VDC

24 VDC BATTERY BANK

DC NEGATIVE BUSBAR

80A MAX DC BREAKER

80A MAX DC BREAKER

DC BREAKER BOX

104F40.EPS

Figure 40   Wiring between batteries and charge controller.

> **NOTE**
>
> The wiring from the batteries to the charge controller will be done later.

7.4.0 Installing the Charge Controller

As with other devices used to make up a photovoltaic system, there are numerous manufacturers of charge controllers. For this section, an Outback FLEXmax™ 60 charge controller will be used as an example. *Figure 41* shows this charge controller.

A charge controller is the first device that actually uses the DC power generated by the panels of the solar array. It is also the first device that makes decisions. Programmable control circuits inside it allow it to monitor the batteries that it is supposed to be charging. It also determines if the batteries are charged or not, and how fast to charge the batteries. Charge controllers also monitor the operations of the inverters that they are

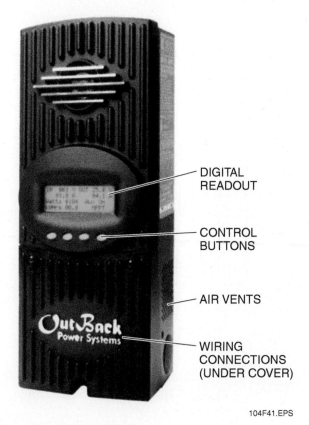

DIGITAL READOUT

CONTROL BUTTONS

AIR VENTS

WIRING CONNECTIONS (UNDER COVER)

104F41.EPS

Figure 41   Charge controller.

feeding PV power. Any PV power not used by the batteries is sent on to the inverter.

Because the charge controllers are so closely related to the batteries and the inverters, they must be installed near each other. Batteries, charge controllers, and inverters all need to be installed in an area that keeps them out of the rain, and out of the sunlight. The FLEXmax™ 60 charge controller only weighs around 12 pounds and can be mounted in different places. For this module, the charge controller will be mounted on a plywood wall. Mounting holes in the back of the controller's housing allow the housing to be attached to almost anything.

7.4.1 Safety

Charge controllers are intended to be installed as part of a permanently grounded electrical system as shown in *Figure 42*.

Because charge controllers are located near batteries, there is always danger from acid fumes or an exploding battery being charged. Because there is some strong DC voltage being processed through the charge controller, there is always the

danger of an electrical shock or possibly even an arc flash.

> **WARNING!**
>
> Working near an FLA battery is dangerous. Batteries generate explosive gases during normal operation. Batteries must be installed in an area that allows air to flow around the batteries. The area also must be ventilated to remove any dangerous gases.
>
> To reduce the risk of injury, charge only deep-cycle lead-acid, lead-antimony, lead-calcium, gel cell, or absorbent glass mat (AGM) types of rechargeable batteries. Other types may explode. Never charge a frozen battery. Always follow proper disposal methods for batteries.

The following are equipment-related restrictions and safety suggestions addressed in the Outback user manual for these FLEXmax™ 60 charge controllers. The restrictions apply unless superseded by local or national codes.

- The negative battery conductor should be bonded to the grounding system at only one point in the system. If a GFCI is present, the battery negative and ground are not bonded together directly but are connected together by the GFCI when it is on. All negative conductor connections must be kept separate from the grounding conductor connections.
- With the exception of certain telecommunication applications, the charge controller should never be positive grounded.
- The ground for charge controller equipment is marked with the ground symbol.
- If a charge controller is damaged or malfunctioning, do not attempt to disassemble or repair it. It must be replaced.
- The charge controller is designed for indoor installation or installation inside a weatherproof enclosure. It must not be exposed to rain and should be installed out of direct sunlight.
- Turn off all circuit breakers, including those to the solar modules, and related electrical connections before cleaning the air vents on the controller.
- The *NEC*® requires ground protection for all residential PV installations.
- All DC battery cables must meet local and national codes.
- Shut off all DC breakers before connecting any wiring to or from batteries.
- Torque all the charge controller's wire lugs and ground terminals to 35 inch-pounds (4 Nm).

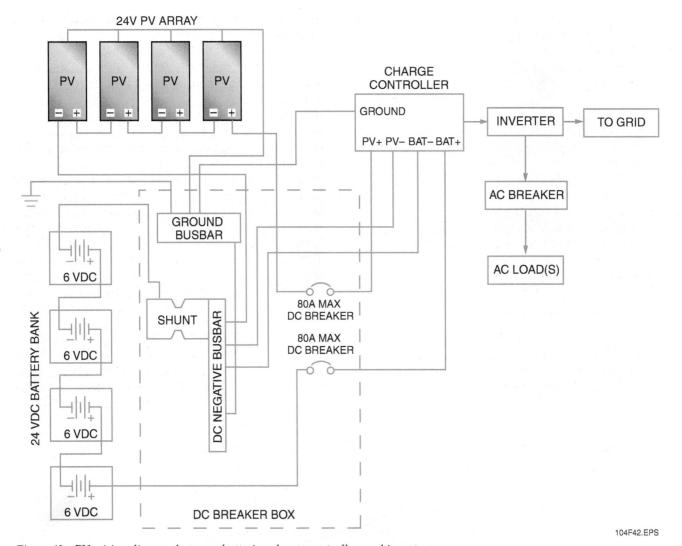

Figure 42   PV wiring diagram between batteries, charge controller, and inverter.

104F42.EPS

- Copper wiring must be rated at 167°F (75°C) or higher.
- Use the appropriate size wire to reduce losses and ensure high performance of the charge controller. Smaller cables can reduce performance and possibly damage the unit.
- Bundle cables using tie wraps.
- Ensure that both cables pass through the same knockout and raceway fittings to allow the inductive currents to cancel.
- DC battery overcurrent protection must be used as part of the installation. Charge controllers can be equipped with either breakers or fuses for overcurrent protection.

Installers must comply with the personal safety guidelines of their individual companies. The following are personal safety suggestions

addressed in the Outback user manual for these FLEXmax™ 60 charge controllers.

- Another installer should remain nearby to provide aid if needed.
- Keep plenty of fresh water and soap nearby in case battery acid contacts skin, clothing, or eyes.
- Wear chemical-resistant eye protection. Avoid touching eyes while working near batteries. Wash hands with soap and warm water when finished.
- Provide a portable eye wash station.
- If battery acid contacts skin or clothing, wash immediately with soap and water. If acid enters an eye, flood the eye with running cool water at once for at least 15 minutes and get medical attention immediately.

- Baking soda neutralizes lead acid battery electrolyte. Keep a supply on hand in the area of the batteries.
- A spill kit is required when working with batteries to contain the acid if a battery leaks.
- Never smoke or allow a spark or flame in vicinity of a battery or generator.
- Be extra cautious to reduce the risk of dropping a metal tool onto batteries. It could short-circuit the batteries or other electrical parts that can result in fire or explosion.
- Remove personal metal items such as rings, bracelets, necklaces, and watches when working with a battery or other electrical current. A battery can produce a short circuit current high enough to weld a ring or the like to metal, causing severe burns.

7.4.2 Charge Controller Installation

The front cover(s) of the charge controller can be removed to expose the mounting holes in the back of the housing. Fasteners (screws or bolts) are inserted through those holes and into or through the material on which the controller is being mounted. In this situation, heavy wood screws inserted through the holes hold the controller housing to a heavy sheet of plywood that is serving as a wall. *Figure 43* shows the mounting holes at the back of the controller's lower wiring section.

Before anchoring the controller's housing to the wall, use a level to ensure that the housing is properly aligned and level. Make sure that there is enough space under the controller for the cabling to enter and exit the housing. The metal raceway that was run down from the combiner needs to be attached into the controller's housing before any wiring can be done. Any metal raceway being run from the charge controller to the battery location also needs to be installed, as does any metal raceway that will be run between the charge controller and the inverter it will be feeding. Refer to local and national codes for guidance about running metal raceway between the charge controller and the batteries and the inverter.

After the controller's housing is properly installed and all the metal raceway has been installed, the wires into and out of the charge controller can be pulled. Be sure to attach only the PV+ wire from the panels (or the combiner) to the

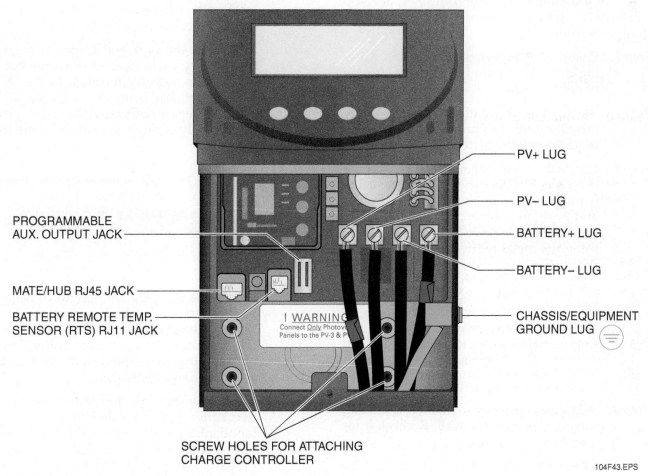

Figure 43 Uncovered lower section of a typical charge controller.

104F43.EPS

PV+ lug. The same goes for the PV– wire and the PV– lug. Farther to the right of the PV lugs are the lugs for the + and – cables that are attached to the batteries. Before attaching the battery cables to the charge controller, make sure that all the cabling between the batteries is secured. Do the same thing for the negative cabling between the batteries and the DC negative busbar, and the negative cabling from the busbar to the controller. On the positive side, make sure that the positive cabling connections from the batteries to the controller are secure. The drawing shown earlier indicated that a DC breaker would be used between the positive side of the batteries and the controller. Check the site design plans to see if a breaker is being used. In this situation, no breaker is being used.

Before installing the charge controller, develop a JSA for this task. Follow these steps to install a charge controller:

Step 1 Ensure that everyone involved is wearing the appropriate PPE for working on or near batteries.

Step 2 Verify that the batteries are properly installed with the proper cabling hook-ups between the batteries.

Step 3 Locate the desired mounting point for the charge controller, and then remove the front covers from the charge controller and position it in its mounting location. Make sure that there is enough room to bring the metal raceway and cables in at the bottom.

Step 4 Align the controller so that it is level, and then anchor it to the wall. Recheck it for alignment and level.

Step 5 Determine where the inverter will be mounted.

Step 6 Attach the metal raceway from the DC disconnect switch to the charge controller.

Step 7 Determine the type and lengths of cable needed to connect the output of the DC disconnect switch to the input of the charge controller.

Step 8 Obtain a fish tape and a small amount of wire lubricant, and then pull the PV+ and PV– wires from the DC disconnect switch to the charge controller, but do not terminate them in the controller.

Step 9 Install all the metal raceway sections that will be connected to the charge controller. Secure the raceway with clamps or other devices as required by code.

Step 10 Determine the type and length of negative battery cable needed from the battery bank's negative terminal to the DC negative busbar being used for the PV system. Prepare and install that cable, but tighten only enough to hold the cable in place.

Step 11 Determine the type and length of negative battery cable needed from the DC negative busbar and the B– terminal on the bottom of the charge controller. Prepare and pull that cable, but do not attach it to the charge controller.

Step 12 Determine the type and length of positive battery cable needed from the battery bank's positive terminal to the BAT+ terminal under the charge controller. Prepare and pull that cable up to the controller, and then loosen the BAT+ lug on the charge controller and slide the end of the positive battery cable into the lug. Tighten the lug to 35 inch-pounds.

Step 13 Prepare the end of the negative battery cable (at the charge controller) for termination, and then loosen the B– lug on the charge controller and slide the end of the negative battery cable into the lug. Tighten the lug to 35 inch-pounds.

Step 14 Verify that the combiner is still de-energized.

Step 15 Determine the type and length of the positive and negative wires needed between the PV system DC safety disconnect switch (located downstream from the combiner) and the charge controller. Pull those wires through the installed raceway.

Step 16 Prepare the ends of the PV+ and PV– wires (from the disconnect switch) for termination. Attach the PV+ wire to the PV+ terminal and the PV– wire to the PV– terminal. Tighten the lug to 35 inch-pounds.

Step 17 Determine the type and lengths of wire needed to connect the DC+ and DC– outputs of the charge controller to the inverter's DC disconnect switch. Do the same for the grounding wires needed from the charge controller to the inverter's DC disconnect switch.

Step 18 Pull and terminate the DC+ and DC– output wires from the charge controller, and tighten the lugs to 35 inch-pounds at the controller terminals.

The next step is to install and wire the inverter.

7.5.0 Installing the Inverter

Like all the other PV system components, there are numerous manufacturers of inverters designed for PV system use. A Sunny Boy 3000US (*Figure 44*) was selected for this training exercise. The one shown is equipped with a DC disconnect switch that allows the installer/maintenance technician to stop the incoming DC power at the inverter instead of going back up to the disconnect at the array.

The Sunny Boy inverter itself weighs about 88 pounds. It is almost 14" tall and 18" wide. It is a little more than 9" thick. The inverter and the DC disconnect switch are both NEMA 3R rated because these inverters may be installed outside

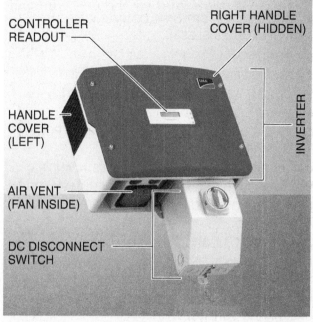

104F44.EPS

Figure 44   Inverter with disconnect switch.

in the weather. For this exercise, the inverter will be equipped with the DC disconnect switch. Both will be installed onto the thick plywood used earlier as a mounting wall.

When the outside cover is removed from the inverter, the terminals and control circuits are visible. *Figure 45* shows the inverter with its cover removed. The chart on the left gives the names or functions of the components marked with the letters.

7.5.1 Inverter Installation

When the Sunny Boy inverter is unpacked, the installer needs to carefully inventory the received parts and make sure that none appear to be damaged. *Figure 46* shows the parts for a Sunny Boy 3000US inverter, the DC disconnect switch, and the mounting hardware needed to install them.

The mounting bracket can be installed on almost any surface, including a standalone post. For better stability, mount the bracket between two wall studs. *Figure 47* shows the mounting bracket installed on a wooden wall.

> **CAUTION**
>
> Inverters such as the Sunny Boy vibrate some as they operate, which is normal. Such minor vibrations or humming can be objectionable if mounted on a wall near a living area.

A – SOCKETS FOR OPTIONAL COMMUNICATION
(RS485 OR WIRELESS)

B – DISPLAY

C – STATUS LEDs

D – OUTPUT AC LINE TERMINALS (N, L1, AND L2)

E – GROUND TERMINAL (PE)

F – PV GROUNDING + DC GROUNDING ELECTRODE
CONDUCTOR

G – OUTPUT AC LINE TERMINALS (L1, L2, N, AND PE)

H – PV GROUNDED TERMINAL (INPUT FROM PV ARRAY)

I – PV UNGROUNDED TERMINAL (INPUT FROM PV
ARRAY)

J – COMBINED UNGROUNDED TERMINAL

K – DC - TERMINAL (INPUT FROM PV ARRAY OR
CHARGE CONTROLLER)

L – DC + TERMINAL (INPUT FROM PV ARRAY OR
CHARGE CONTROLLER)

M – FLAT CONNECTION FOR GROUNDING THE CABLE
SHIELD FOR COMMUNICATION

N – TERMINAL FOR OPTIONAL COMMUNICATION (RS485)

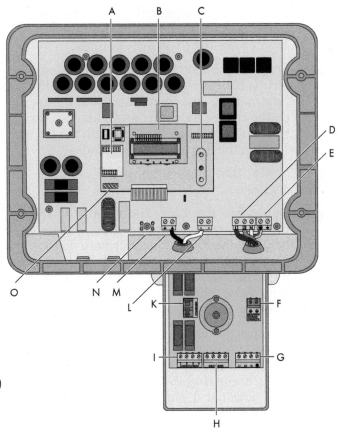

104F45.EPS

Figure 45 Inverter components.

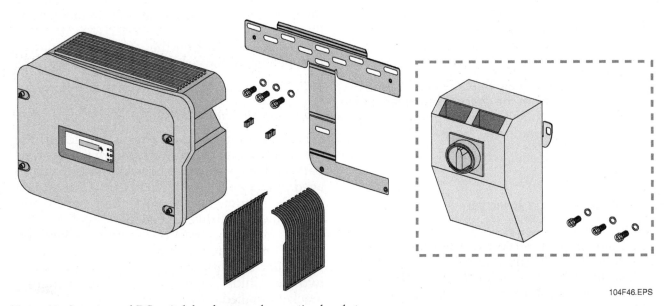

104F46.EPS

Figure 46 Inverter and DC switch hardware and mounting brackets.

After the mounting bracket is installed, the DC switch is attached to the mounting bracket. Knockouts in the bottom and sides of the DC switch allow the raceway and wiring to be attached. Those wires will feed up and into the inverter when it is installed. *Figure 48* shows the DC disconnect switch installed and the wiring ports on the bottom of the inverter.

Figure 49 shows the inverter itself being hung onto the mounting bracket. After the inverter is hung, the mounting screws are installed, and then the handle covers are attached.

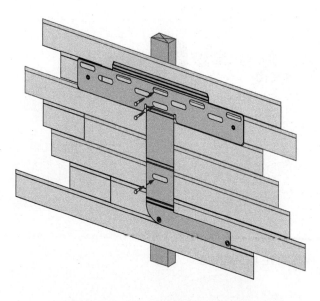

CENTERED ON A SINGLE STUD

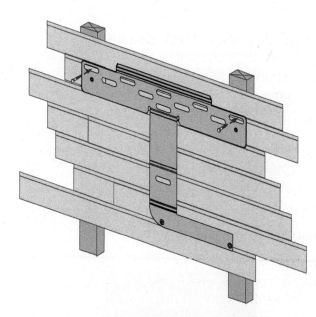

CENTERED BETWEEN STUDS

104F47.EPS

Figure 47   Mounting bracket on a wooden wall.

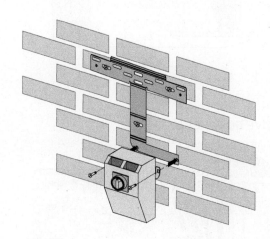

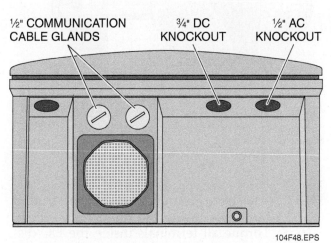

½" COMMUNICATION
CABLE GLANDS   ¾" DC   ½" AC
                     KNOCKOUT   KNOCKOUT

104F48.EPS

Figure 48   Installed DC disconnect switch and location of
inverter wiring ports.

WARNING!

To prevent personal injury or damage to the
equipment, get someone else to help mount the
88-pound inverter.

Before installing the inverter, develop a JSA for
this task. Follow these steps to install a Sunny Boy
3000US inverter with DC disconnect switch:

Step 1   Position, level, and install the inverter
mounting bracket as indicated on the
site design plan. Make sure that it is
close enough to mate with the metal
raceway and the wiring from the charge
controller.

Step 2   Position, level, and install the DC disconnect switch as indicated on the site
design plan.

Step 3   Verify that there is no power coming
in on the PV+ and PV– wires from the
charge controller.

Step 4   Thread the PV+ and PV– wires (and
grounding wires) from the charge controller into the disconnect switch housing, and then attach the raceway from
the charge controller to the body of the
DC switch. *Figure 50* shows a simple
wiring diagram of a Sunny Boy inverter
equipped with a DC disconnect switch.

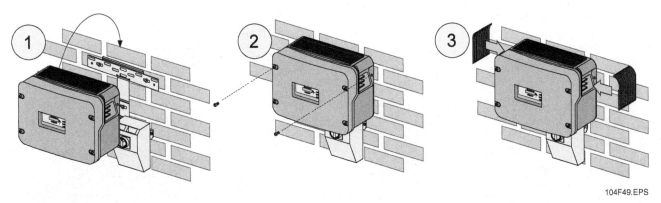

Figure 49 Attaching an inverter.

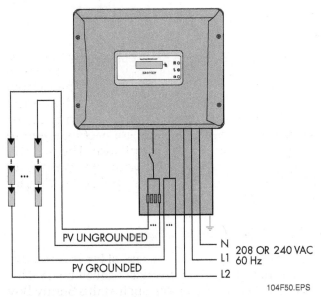

Figure 50 Inverter wiring diagram.

PV UNGROUNDED

PV GROUNDED

N
L1 208 OR 240 VAC
L2 60 Hz

104F50.EPS

Step 5 Pull the PV and grounding wires into the DC disconnect switch housing. The inverter must be connected to the AC ground from the building electrode ground.

> **NOTE**
>
> The PV array (frame) ground should be connected to the PV grounding and the DC grounding electrode conductor. The size of the PV grounding conductor is usually based on the size of the largest conductor in the DC system. A DC grounding electrode conductor may be required by local code.

Step 6 Attach the PV+ (positive DC) lead(s) to the *ungrounded* terminals inside the DC switch, and attach the PV– (negative DC) to the *grounded* terminals (*Figure 51*).

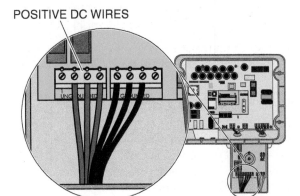

POSITIVE DC WIRES

NEGATIVE DC WIRES

GROUNDING TERMINALS

104F51.EPS

Figure 51 Attaching PV+ and PV– wires to DC disconnect switch.

Step 7 Attach the grounding wires to the ground terminals inside the disconnect switch. Torque all terminals inside the DC disconnect switch to 15-foot-pounds.

Step 8 Identify the wiring that will be used to connect the DC disconnect switch terminals to the inverter terminals. The DC switch may or may not be already wired.

> **NOTE**
>
> Some inverters have a rubbery grommet designed to mate with the attached raceway. A membrane seal across the opening of the grommet has to be punctured for the wires to enter the inverter housing. When the wires are pushed through, the membrane forms a seal around the wires.

Step 9 Size and install the short section of raceway needed to connect the inverter to the output port of the disconnect switch.

Step 10 Pull the wires that will connect the switch to the inverter through the installed raceway.

> **WARNING!**
>
> Remember that this inverter weighs 88 pounds. An assistant is needed to help lift, align, and set the inverter. The assistant can also help start and tighten the mounting screws.

Step 11 Position the inverter above the DC disconnect switch and slide the inverter housing down onto the mounting bracket, while at the same time mating the DC switch raceway to the inverter raceway connector. Install and tighten the mounting screws and raceway connectors.

> **NOTE**
>
> Due to the closeness of the DC switch top to the bottom of the inverter, the wires or cables from the DC switch have to be pulled up into the inverter as the inverter is being mounted.

Step 12 Separate the PV wires that were pulled up from the DC disconnect switch, and then connect the black PV+ (positive DC) and white PV– (negative DC) wires to their respective terminals inside the inverter (*Figure 52*). Torque all inverter terminals to 15 foot-pounds.

At this point, any DC voltage fed from the PV panels and charge controller would be sent into the DC-to-AC conversion circuits inside the inverter. On this particular Sunny Boy inverter, the AC terminals are at the lower-right corner

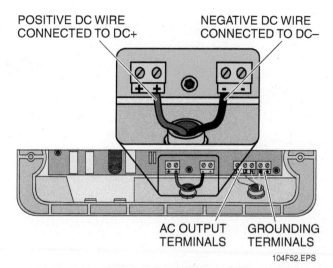

POSITIVE DC WIRE CONNECTED TO DC+   NEGATIVE DC WIRE CONNECTED TO DC–

AC OUTPUT TERMINALS   GROUNDING TERMINALS

104F52.EPS

Figure 52   PV+ and PV– terminal connections.

of the inverter. When this inverter is being used with a DC disconnect switch, wiring must be installed from the inverter's AC terminals down to the AC terminals inside the DC disconnect switch housing. From there, the AC wiring can be sent on to the circuit breaker connecting either to a grid circuit or to an AC appliance. The inverter still must be programmed with the specifics of how it is to be used, but at this point, the inverter is installed.

7.5.2 AC Hookup

As noted earlier, the AC output of the inverter can be tied into a grid connection or into an off-the-grid AC breaker. Inverters, such as this Sunny Boy 3000US inverter, may or may not be tied through a DC disconnect switch. Since the previous section had the inverter tied through such a switch, this discussion will continue along those lines. *Figure 53* shows a wiring diagram of a Sunny Boy inverter equipped with a DC disconnect switch whose outputs feed through an AC disconnect switch. From the AC disconnect switch, the AC output goes to an AC distribution panel tied to the utility.

In *Figure 53*, the DC power from the PV array is being fed directly into the DC inputs of the inverter. This setup has no DC disconnect near the panels, and no charge controller or batteries. This particular drawing does have the array power coming through the left side of a DC disconnect switch and into the inverter. The AC output from the inverter is being fed down through the right side of a DC disconnect and over to an independent AC disconnect switch. From there, the AC power goes on to the right and ties into a 30A breaker installed in the main disconnect receiving power from the utility. When tied to a

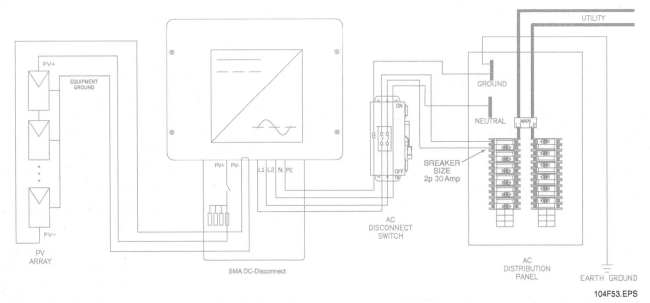

Figure 53 Inverter's AC feed to a grid-tied AC distribution panel.

grid, the control circuits inside the inverter monitor the grid and make the necessary adjustments to produce the proper AC output. If the grid goes down, the inverter should stop working.

For this exercise, the PV system will be set up for off-the-grid service. The AC output from the inverter will go through the DC disconnect to an AC breaker. For training purposes, that AC breaker output will be tied to an AC appliance. An AC disconnect switch could be used, but will not be since the output will be going to an AC breaker that can be used to kill the AC power to

the AC appliance. *Figure 54* shows a simple block diagram of a PV power distribution that can be set up for training purposes.

When wiring the output of the PV system to the AC breaker, pull the AC wires that will connect the two devices. At that point, the individual wires can be attached. *Figure 55* shows a closer view of the AC wiring terminals on the DC disconnect switch used with a Sunny Boy inverter.

Before making wiring terminations, develop a JSA for this task. Refer to *Figure 55* and follow these steps to terminate the AC wiring between

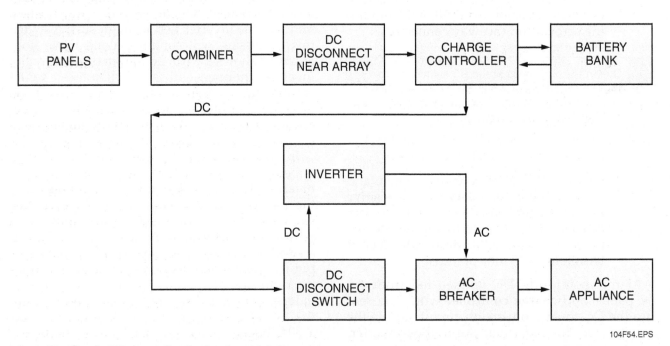

Figure 54 Simple PV power distribution for training purposes.

NCCER — *Contren® Learning Series*   57104-11

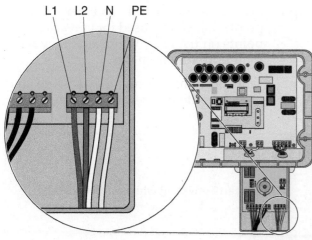

L1   L2   N   PE

104F55.EPS

Figure 55   AC wiring terminals on a DC disconnect switch.

the AC breaker and the DC disconnect switch used with a Sunny Boy inverter:

Step 1   Verify that the AC breaker is opened and inserted into the bus inside its enclosure.

Step 2   Determine how much AC wiring will be needed to run wire from the circuit breaker enclosure back to the output terminals of the DC disconnect switch. Allow an additional 4'.

Step 3   Pull the AC wiring that will be used between the AC circuit breaker enclosure and the DC disconnect switch.

Step 4   Connect the AC equipment ground wire to the PE terminal inside the DC disconnect switch and tighten until snug.

Step 5   Connect the L1 (AC line 1 or the Ungrounded) wire to the DC disconnect switch terminal shown as L1. Tighten until snug.

Step 6   Connect the L2 lead (AC line 2) to the DC disconnect switch terminal shown as L2. Tighten until snug.

Step 7   Connect the N (AC line N) to the DC disconnect switch terminal shown as N. Tighten until snug.

Step 8   Identify the jumper wires needed to tie the AC output terminals of the DC disconnect switch to the AC input terminals of the inverter, and attach them to the inverter terminals.

Step 9   Torque all terminals in the inverter and DC disconnect switch to 15 foot-pounds.

Step 10   Locate the ground busbar in the circuit breaker enclosure, and connect the PE wire.

Step 11   Locate the neutral busbar in the circuit breaker enclosure, and connect the white N wire to it.

Step 12   Connect the black AC wire to the input side of the circuit breaker.

Step 13   Wire an AC outlet to the output of the AC breaker so that an AC appliance such as a fan can be attached.

Step 14   Torque all terminals to 15 foot-pounds unless told to do otherwise.

Step 15   Locate the inverter cover and all the mounting screws and locking washers.

Step 16   Inspect the inverter cover for damage, especially on the sealing area. If the seal is bad, the cover will not keep the moisture and dirt away from the sensitive control circuits inside the inverter.

Step 17   Replace the inverter cover, making sure to replace and tighten all mounting screws and locking washers.

At this point, the solar panels can be installed and connected to the PV system components. Remember that DC voltage will start flowing into the PV components as soon as the PV panels are wired into the system.

7.6.0 Installing the PV Panels

As noted earlier, the PV panels produce DC voltage. When tied by wiring into the PV system at the combiner, that DC voltage starts to flow. If the DC disconnect switch near the combiner(s) is closed, that DC voltage flows down to the charge controller. From there, it flows on to the batteries and the inverter. This section deals with the installation of the PV panels onto the mounting hardware that was installed earlier.

In an earlier section, instructions were given for installing the beams for the solar panels. Now, the solar panels must be transported onto the roof and secured to the beams. For training purposes, only four panels will be installed. They will be installed vertically (up and down a sloping roof). Two panels, arranged parallel to each other, will be on the top row. They will be tied in series. Two more arranged in the same manner will be on the bottom row. The two rows of panels will be arranged electrically parallel to each other and fed into a single combiner. As seen from the gutter edge of the roof, the combiner will be mounted on the left of the array near the center of the top and bottom rows of panels. *Figure 56* shows how the panels should appear after installation. Notice the location of the end and mid clamps.

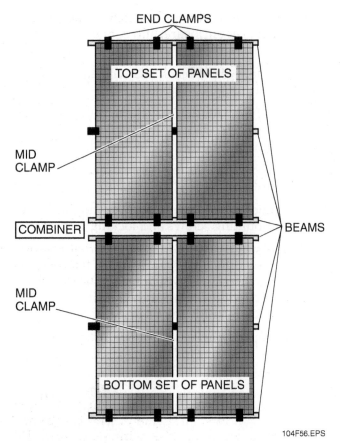

END CLAMPS

TOP SET OF PANELS

MID CLAMP

COMBINER

BEAMS

MID CLAMP

BOTTOM SET OF PANELS

104F56.EPS

Figure 56   Layout of panels and clamps.

Before installing the solar panels, develop a JSA for this task. Follow these steps to install solar panels:

Step 1   Assemble the solar panels, installation hardware, and tools needed to secure the panels to the mounting beams. Remember that the bolts must be torqued to the manufacturer's specifications, typically 75 foot-pounds. Take only the wrenches and sockets needed to tighten and torque the bolts and nuts being used.

Step 2   Verify that all personnel are wearing the appropriate PPE and fall protection for roof work. Also make sure that any ladders or scaffolding are properly placed and have been inspected prior to use.

> **NOTE**
>
> If an aerial lift must be used, verify that its operator is properly trained and certified.

Step 3   Inspect the lifting device that will be used to transport the materials to the roof. Operate it without a load to make sure it works.

Step 4   Locate or install all necessary anchor points that will be used for lanyard tie-off points on the roof. Verify that they meet minimum requirements.

Step 5   Transport the clamping hardware and tools up onto the installation site.

> **WARNING!**
>
> Do not carry anything while climbing ladders.

Step 6   Assemble all the sliders and slide them into their approximate mounting locations on the beams.

Step 7   Position the first solar panel so that it is near the top-right corner of the mounting beams. Align the panel so that it is evenly positioned across the top three mounting beams.

> **NOTE**
>
> Small, lightweight C-clamps would be handy for temporarily holding the panel in place until its position can be verified, and the permanent clamps can be installed.

Step 8   Move the sliders so that their bolts are positioned at the desired locations for the end and mid clamps, then install the end and mid clamps. Tighten the nuts hand tight.

Step 9   Verify again that the panel is properly aligned on the beams, and that all its clamps are properly positioned, then use a torque wrench and socket to tighten the locking nut for each clamp to the desired torque.

Step 10   Install the second top panel in the same manner as the first, but leave a 1" space between the panels for the mid clamps. Install all clamps and torque all nuts and bolts.

Step 11   Connect the wiring of the top two panels, in series, and run their wires over to the input wires to the combiner. Do not connect the panel wires to the combiner wires. *Figure 57* shows panels wired in series.

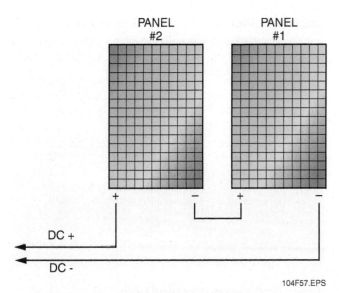

Figure 57 Panels wired in series.

Step 12 Position the first bottom row panel so that its termination box is toward the top and the panels already installed. Align it and secure it.

Step 13 Install the second bottom row panel in the same manner as the first bottom row panel, making sure to leave the 1" space between the panels for the mid clamps. Tighten the nuts and bolts to the required torque.

Step 14 Attach and secure all grounding devices to the panels as required by code.

Step 15 Connect the DC wiring between the two bottom panels so that they are in series.

Step 16 Run the output wires from bottom string of panels over to the combiner, but do not connect them to the combiner input wires.

Step 17 Open the combiner cover and determine which of the extender wires go to which combiner circuit breakers. Label or mark the DC+ and DC– as wire sets 1, 2, 3, and so on.

> **NOTE**
>
> In this training situation, only two sets of wires will be needed since there will be only two strings feeding into the combiner. Masking tape and a permanent marker can be used to temporarily bundle and mark the different sets of wire.

Step 18 Move the On/Off switch of the DC disconnect switch to the Off or open position. This will keep any DC power from the panels from continuing down the PV system until the panels are all installed.

Step 19 Install all the cable clamps required by code to secure the wiring between the panels and the combiner. Loose wires are unacceptable.

> **WARNING!**
>
> When the DC wiring from the panels is connected to the DC input wiring to the combiner, DC voltage from the panels will be present at the combiner terminals. That voltage is potentially dangerous. From this point on, treat all wires, connectors, and terminals as if they are energized.

8.0.0 INSTALLING, LABELING, AND TERMINATING WIRING PER CODE

Up to this point, the PV equipment has been installed, but not completely wired together. Installers terminating the wiring between PV components must be qualified electricians knowledgeable of all current electrical and building codes applicable to the installation site. Before the panels are wired into the combiner, and the disconnect switches are all closed, all the circuits must be labeled and methodically checked before being totally activated.

8.1.0 Labeling

The labeling process includes a complete set of as-built drawings, photos, and schematics so that anyone working on the system years later will know where each component is located, and what it should be doing. All this information should be in the system design plans for the site. The following are some of the labeling required by the *NEC*®:

- *Ground-fault protection device* – Labels and markings must be applied at a visible location near the device stating that if a ground fault is indicated, the normally grounded conductors may be energized and ungrounded.
- *Bipolar source and output circuits* – The equipment must have a label that reads: "Warning – Bipolar Photovoltaic Array. Disconnection of Neutral or Grounded Conductors May Result in Overvoltage on the Array or Inverter."

- *Single 120V supply* – If a single-phase three-wire distribution panel is fed with a 120V PV inverter output by connecting a jumper wire between the panel busbars, the panel must have a label that states: "Warning – Single 120-Volt Supply. Do Not Connect Multiwire Branch Circuits!"
- *System disconnects* – Each PV disconnecting means in the system must be permanently marked to identify it as a PV system disconnect. Note that disconnects may not be located in bathrooms.
- *Energized terminals in open position* – In cases where both terminals of a device may be energized, even if the device is in the open position, a warning sign must be mounted on or adjacent to the disconnecting means that states the following or equivalent: "Warning! Shock Hazard. Do Not Touch Terminals. Terminals On Both The Line And Load Side May Be Energized In The Open Position."
- *Photovoltaic power source* – A label must be installed near the PV disconnecting means that states (1) the operating current, (2) the operating voltage, (3) the maximum system voltage, and (4) the short circuit current.
- *Identification of power sources, stand-alone systems* – A permanently installed plaque or directory must be installed outside the building to indicate the locations of all system disconnecting means, and to indicate that the structure contains a stand-alone electrical power system.
- *Identification of power sources, facilities with utility connections* – A permanently installed plaque or directory must be installed showing the location of the service disconnecting means and the PV disconnecting means.
- *Busbar or conductor connection equipment* – Any equipment containing an overcurrent device that supplies power to a busbar or conductor must be marked to indicate the presence of all sources. For example, if the point-of-utility connection for an inverter is at the main disconnect device or meter of a building, there may be a fused or circuit breaker disconnect near the main disconnect. If so, it needs to be labeled.

Most wire labels are the same as any other wiring labels found in conventional electrical panels. The labeling on fuses and circuit breakers may be more specific to the PV equipment. *Figure 58* shows the wires attached to the circuit breakers and a terminal strip.

Check state and local requirements for labeling. If the PV system is being tied into a utility, check the utility labeling requirements. If batteries are used, local fire inspectors will inspect the battery

BREAKER LABELS   WIRE LABELS

WIRES INTO AND OUT OF BREAKER   104F58.EPS

Figure 58   Labeling on breakers and wiring.

disconnect. Per *NEC Section 702.7*, the service-entrance equipment must be labeled to indicate the type and location of on-site optional standby power sources.

The labels attached on the outside of the enclosures are often made by the installer company, or they are bought from companies that make and sell generic labels for PV equipment. *Figure 59* shows samples of labels found on the outside of PV enclosures.

8.2.0 Checking Terminations

When putting the finishing touches on the PV wiring, an installer needs to know what the voltage levels and the current levels should be as the components are tied together. The system design plans should have those details, but actual measurements are the only way to validate the numbers on the plans. There will normally be a small amount of voltage drop as the longer wires are connected, but the system designer should have already made adjustments for such drops. When a voltage reading is taken between the DC+ and DC– leads from the panels, the voltage is referred to as being open-circuit voltage because the leads have no load on them at that point. When a current reading is taken at that point, it is referred to as being short-circuit current. The open-circuit voltage values and short-circuit current values are also measured at the combiner and other PV components down the line. Until an actual load is placed on the PV system, all electrical values will be open-circuit voltage values and short-circuit current values. The open-circuit voltages should normally remain within two percent of each other as long as the irradiance (sunlight) on the panels

Service Labeling

Labels are installed near the metering equipment to alert fire officials and utility personnel to the presence of another power source.

104SA02.EPS

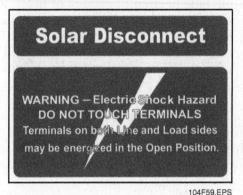

104F59.EPS

Figure 59 Examples of labeling for exterior of PV enclosures.

stays the same during the testing. The short-circuit currents should be within five percent of each other during the test.

The control circuits in the charge controller(s) and the inverter(s) must be programmed and adjusted to the voltages and currents of the system in which they are being installed. Most charge controller and inverter control circuits are programmed to a default setting at the factory, but they need to be reprogrammed on site for each site's application. To fully activate and test a PV system, installers must first finalize all the

terminations, and then document all the voltages and current levels throughout the system.

8.2.1 Panel Terminations and Outputs

As a rule, most solar panels today come prewired with a termination box or junction box on their backside. The wires from the modules inside the panel should already be terminated inside the box. The box should have a DC+ wire and a DC− wire coming out of it. Those output wires should have a male and a female MC connector on them. The design of those MC connectors should prevent accidental contact with the electrical wires, but be careful when handling these connectors. A multimeter, set up to measure DC voltage, can be used to measure the output voltage from individual panel leads. Since the DC voltage is generated by the solar panels, the voltage checks must start at the output wires of the panels. Refer to the nameplate information on the back of a panel to see exactly what voltages the panel should be producing when exposed to full sunlight. Change the meter and take a current reading. Record the values found.

8.2.2 Combiner Terminations

Before coupling the MC connectors of the solar panel output wires to the corresponding wires going into the combiner, check the termination of the combiner wires. Make sure that they are properly terminated and properly labeled. If the combiner is built to handle the inputs from six strings of panels, the incoming wires should be labeled 1 through 6. If fewer inputs are needed, fewer input wires should be terminated inside the combiner. If the array has only four strings of panels, the outputs of those strings should be connecting to the first four input wires of the combiner. The terminals for those inputs will be at the input side of circuit breakers or fuses. Review the torque specifications for the combiner being used and make sure that the terminal nuts and bolts holding the input wires are torqued to the proper level. When checking torque, loosen the nut slightly and then reset the torque. If the nut is not loosened before torque is checked, the torque wrench will put more torque than desired onto the nut and bolt. After repeated torque checks (without loosening the nut a little), the level of torque exceeds what the nut and bolt threads can tolerate. The result is a failure.

As for checking the voltages inside the combiner, there should be no voltage until the panel output wires are connected to the combiner input wires. When connected, the DC voltage seen at the panel output wires should be the same at the corresponding terminals inside the combiner. If a combiner is equipped with a circuit breaker for each of the DC+ inputs, that breaker can be opened to ensure that there is no backfeed voltages from any other circuits. With the breaker opened, the voltage measured on its input should be the same as the voltage measured at the panel output lead.

When all the breakers are closed in the combiner, the input voltages should combine to create one larger voltage on the output leads of the combiner. If two 6V panels were tied in series to form a string, that string should produce 12 VDC, which would be felt at the output of the combiner. When two more 6V panels are wired in series to form a string, and that string is connected in parallel to the first string, the combiner sees the two strings as two parallel voltage dividers. The multimeter should still read 12 VDC because of the parallel circuits. If an amp meter had been placed on the output of the combiner when only one string was providing power, the current level would have been at a given amount. If that amp meter was left in place as the second string of panels was added, the current level should have increased. If a third string of panels was added later, the current level would increase again, but the voltage level would stay the same. Remember that solar panels are wired in series to increase their overall output voltage. They are wired in parallel to increase their overall level of current. When panels are wired in series to form a string, and multiple strings are aligned to run parallel into a combiner, the overall voltage and current of the array is increased. A normal pattern in PV systems is to run four panels in series to form a string producing a desired voltage level (24 VDC or 48 VDC total), and then run multiple strings to get a desired current level. The use of more than four panels per string is not recommended. The dangers of electrical shock increase as additional strings of panels are combined to obtain those higher voltage and current levels. As the voltages and currents are checked, record the findings.

> **WARNING!**
> To prevent personal injury caused by electrical shock, treat all wire ends and terminals as if they are potentially dangerous. Use a multimeter to verify that any circuit is safe.

The outside of a combiner should have some kind of nameplate indicating that there is always PV power coming into the combiner's input terminals. The wiring into and out of the combiner terminals must be labeled as to where they are coming from and where they are going.

8.2.3 DC Disconnect Switch Terminations

Most codes require that a DC disconnect switch be placed between the panels of an array and the rest of the PV components. Some codes require that the disconnect switch be near the ground and others require it to be near the panels. The reason is safety. If any work must be done on the charge controller, inverter, or the wiring between them, the DC power from the panels must be stopped. DC disconnect switches receive DC power from the combiner. The wiring is heavier due to the higher levels of current being carried, but the input and output wires are still terminated at lugs inside the switch enclosure.

If the panels have been connected to the combiner already, and the combiner breakers are closed, there will be DC power on the input wires to the disconnect switch. To prevent electrical shock, open the combiner breakers before checking the torque on the input and output terminals of the DC disconnect switch. After the torque is checked, close the combiner breakers and make the electrical checks.

After checking the wiring terminations, check the mechanical movements of the switch contacts. When satisfied that the mechanical parts of the DC disconnect switch are operational, make sure that the downstream components (charge controller and inverter) are ready for the DC power from the panels. When all that has been checked, allow the DC power from the combiner to enter and exit the disconnect switch. Make voltage and current checks on the input and output lines of the switch. As the electrical values are checked, record the findings. When the switch is closed, the charge controller and batteries becomes a load to the voltage being supplied from the panels.

8.2.4 Charge Controller, Battery, Inverter, and AC Breaker Terminations

A charge controller has multiple inputs and outputs. It receives DC power through the DC disconnect switch and the solar panels. It produces DC outputs to both the batteries and to at least one inverter. It monitors the charges on the battery bank, and the temperatures of the batteries as they charge. It also monitors the inverter(s) that it feeds.

Before attempting to check the wiring terminations and torque values on the charge controller and the inverter, open the DC disconnect switches near the output of the combiner and at the input to the inverter. Also, open the AC breaker at the output of the inverter. If there is a disconnect switch between the charge controller and the battery bank, open it. After the disconnect switches are opened, use a multimeter to verify that there is no voltage on the terminals of the charge controller and the inverter. Review the torque specifications for all the terminals and then check the torque on all the charge controller, inverter, and battery terminals. Also, ensure that all the labeling on wiring and components inside the enclosures meet code, and then check the exterior labels.

After the terminals and labeling have been checked, check the wiring to and from the batteries. Make sure all the cabling polarities are right.

Verify that the batteries have been prepared to receive a charge. Review the system design plan and the user manual for the charge controller to determine how the batteries will be charged. The control circuits in the charge controller may need to be changed from their default settings. Refer to the user manual for instructions on how to change any controller settings.

After the controls for the charge controller have been set for charging the battery bank, and the batteries have been checked, check the inverter controls. Refer to the system design plan and the user manual for the inverter to set up the inverter controls. When the inverter is ready, check the wiring and terminals connecting the inverter to the AC circuit breaker. If all those are good, close the DC disconnect switches and start making all the electrical checks. Keep the AC breaker open until the output from the PV system is ready to be used. As the electrical checks are made, record the findings.

9.0.0 ACTIVATING AND TESTING PV SYSTEM TO VERIFY OPERATIONS

Before any PV system is turned over to the customer, it must go through a final inspection and then put under a load to verify that the system operates as planned. This final phase is broken down into three activities: final inspection, activation (placing it under load), and testing to see how it reacts under load.

9.1.0 Final System Walkdown

The PV system was inspected earlier as the wiring and terminals were checked, but one more inspection is needed before it is activated and tested under a load. Start again at the panels and work down to the AC breaker. Remember that these inspections are being made on equipment with live electrical circuits. Comply with all electrical safety guidelines applicable to working on or near electrical voltage. Before inspecting the installed system, develop a JSA for this task. Follow these steps to make a final walkdown of the PV system:

Step 1   Verify that all personnel are wearing the appropriate PPE and required fall protection equipment. Also make sure that any ladders or scaffolding are properly placed and have been inspected prior to use.

Step 2   Gain access to the underside of the installation area and verify that all anchoring devices for the mounting beams appear to be properly installed.

Step 3 Locate all grounding electrodes and verify that the wiring is undamaged and secure.

Step 4 Inspect the grounding wires from the grounding electrodes back to the individual PV components. Verify that the wire is undamaged, securely anchored, and properly attached as required by code.

Step 5 Gain access to the roof and anchor the fall protection equipment.

Step 6 Verify that the standoffs and mounting beams are properly installed at the designed height above the roof surface.

Step 7 Verify that the solar panels are properly aligned to the mounting beams, and that all mounting clamps are properly installed. Also, verify that the panel surfaces are clean.

Step 8 Verify that all DC+ and DC– wiring from the panels to the combiner is properly connected, secured (*Figure 60*), and labeled in accordance with all codes.

> **NOTE**
>
> It is extremely important that installers keep the cabling neatly bundled and securely fastened. These cables are exposed to all weather conditions. Flopping cables become damaged over time. They also irritate anyone hearing them slap or rub against the surface of the roof or against any of the supporting rails.

WIRING CLIP

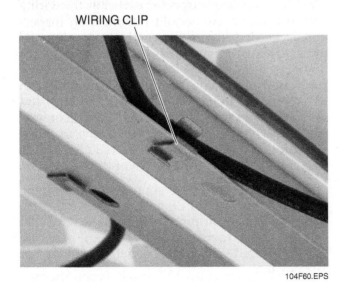

104F60.EPS

Figure 60 Wiring clip.

Step 9 Verify that the combiner is securely fastened and labeled, and that all wiring and terminals inside it appear undamaged and functional.

Step 10 Verify that the panels and their mounting beams are properly grounded.

> **WARNING!**
>
> Do not attempt to climb down a ladder while carrying anything.

Step 11 Disconnect fall protection and patch the attachment holes in the roof.

> **WARNING!**
>
> Exercise extreme caution after disconnecting fall protection devices.

Step 12 Verify that the charge controller is securely mounted and labeled inside and out.

Step 13 Verify that the charge controller wiring and terminals appear to be securely fastened and properly labeled.

Step 14 Check the batteries and their cables back to the charge controller. Verify that the batteries appear to be charging properly.

Step 15 Refer to the charge controller user manual and system design plan to see how the controller should be programmed.

Step 16 Verify that the inverter appears to be undamaged and that its wiring and terminals appear to be securely fastened and properly labeled.

Step 17 Refer to the inverter user manual and system design plan to see how the inverter should be programmed. Verify that the inverter is operating correctly.

Step 18 Verify that the raceway between components is undamaged and securely fastened.

Step 19 Verify that the AC breaker and its enclosure appear to be undamaged and that its wiring and terminals appear to be securely fastened and properly labeled. Also, verify that the breaker is properly labeled inside and out.

9.2.0 System Activation and Testing

If all the checks made during the final walkdown were good, the system can be activated. If the PV system is going to be used only to power standalone AC appliances, the AC breaker can be closed and the connected appliance(s) should operate. If the PV system is going to be tied into a grid-fed service connection (main breaker panel) to a home or business, additional precautions need to be met to comply with the utility company (grid) guidelines for connecting PV systems. For this training, the PV system will be a simple standalone system powering at least one AC appliance.

Up to this point, the charge controller has been using PV system power to charge the batteries. That process was seen as a load to the PV power supply. When the AC breaker is closed, any connected AC appliance will present an additional load to the PV system. After the system has been loaded, final operational checks can be made.

> **NOTE**
>
> Since each system will be different due to different PV components being used, this module cannot address specific voltage and current levels. Solar installer instructors or trainers setting up their own equipment for any tests will need to come up with the electrical values for their specific systems.

An acceptance test needs to be performed under ideal test conditions. Those test conditions are normally called standard test conditions or STC. Under standard test conditions, the array should be receiving full sun (1,000 watts per square meter of panel space) and the temperature on the surface of the panels should be 131°F (55°C). To get a true reading on the overall PV system, the system needs to be checked without any battery storage system attached. Doing so simplifies the process by removing a major load from the system. The goal is to determine how much actual power can be expected at the output of the inverter.

The following is an example of an STC test performed on a 2 kW system with crystalline silicon panels feeding a utility-interactive inverter. No battery storage was being used. Even under ideal testing conditions, some of the power produced by the panels was lost within the overall PV system. Power is lost in the panels themselves, but it is lost primarily in the wiring and the PV components (combiner, switches, charge controller, and inverter). Here are the calculations that were used to determine the losses throughout the system:

1. Change all loss figures to consistent numbers. For example, if a loss of 10 percent is expected, then an efficiency factor of 90 percent applies to that loss. So for a 10 percent loss, the number 0.9 represents the 90 percent efficiency factor.
2. Multiply all the numbers obtained in the first step of this exercise together to get the overall efficiency factor, expressed as a decimal.
3. Multiply the starting amount by the overall efficiency factor to obtain the remaining amount, or output.

The procedure for the example gives the following results:

- Efficiency factor for module tolerance, array mismatch, and dust degradation = 0.9
- Loss from array temperature degradation: $(55 - 25) \times (-005) = -0.15$, which represents a percent loss expressed as a decimal. Efficiency factor 0.85.
- Efficiency factor from 3 percent wiring losses: 0.97
- Efficiency factor from 6 percent inverter losses: 0.94
- The output, taking into account all four loss mechanisms: $2,000 \times 0.9 \times 0.85 \times 0.97 \times 0.94 = 1,395$ watts.

According to those testing the 2 kW system, the results they found (1,395 watts) was consistent with actual field finds. As a result of their testing, a general rule of thumb was determined. It stated that the output of a utility-interactive system with no storage can be expected to be about 70 to 80 percent of the sum of the rated power of the panels. The 1,395 watts that was calculated for the 2 kW system was close to the 70 percent value of the expected output of the panels. When weather conditions change those STC specifics, that percentage number will also change. The same goes for when the irradiance decreases.

Before activating a PV system, develop a JSA for this task. Follow these steps to fully activate a PV system for acceptance testing.

Step 1 Determine the best time of day to activate the complete PV system, and then schedule the test.

Step 2 To unload the PV system, open the disconnect switch at the combiner to stop power from flowing downstream to the charge controller. Also, open the AC breaker on the output of the inverter.

Step 3 Use a multimeter to verify that charge controller is producing no output voltage to the batteries.

Step 4 Disconnect the charge controller output cable at the input to the batteries. Secure the cable to make sure that it is not touching anything after being disconnected.

Step 5 Use the multimeter to verify that the inverter has no output voltage.

> **NOTE**
> At this point, the output from the array panels should be going through the combiner and stopping at its DC disconnect switch. When the switch is closed, the PV power should continue on into the charge controller. With no battery connection, the charge controller should be sending the PV power directly on to the inverter.

Step 6 Determine what the output of each solar panel being used should be under the given test conditions. If two or more panels are wired in series to form a string, determine what the output of the string should be under the conditions.

Step 7 Close the DC disconnect switch at the output of the combiner and then measure and record the open-circuit voltage value at the following points:
- Output of each string of panels
- Output of the combiner
- Input of the charge controller (to see how much is lost through the cabling from the combiner down to the charge controller)
- Output of the charge controller
- Output of the inverter

> **NOTE**
> Because of the charge controller, this process is a little different than the test that was run on the 2 kW system, but the ideas are the same.

Step 8 Determine how much loss is happening across the PV system.

Step 9 Determine how much actual PV power is reaching the output of the inverter.

Step 10 Determine if the measured output from the inverter meets the system design plan for the given PV system.

Step 11 Open the DC disconnect switch at the combiner and verify that the charge controller output is back to zero volts.

Step 12 Reconnect the cable from the charge controller to the batteries.

Step 13 Check to see if the batteries are fully charged. If they are, move on to the next step.

Step 14 Change the charge controller so that the batteries are supplying power to the inverter.

Step 15 Measure the PV voltage and current at the output of the inverter to see how much of a difference there is between power from the array panels versus power from the batteries. Expect the inverter output to be about 4 percent less with battery power.

> **NOTE**
> If everything is working at this point, the system should be turned over to the customer. Most companies use an acceptance test checklist.

Step 16 Clean the work area and store all tools and equipment used in the installation.

Step 17 Complete all required paperwork to close out the job.

Summary

System installation begins with a careful review of the site assessment and system design plans. Next, site safety hazards must be assessed and a job safety analysis prepared for each task. After the plans have been reviewed and the safety analysis is complete, the materials are inventoried and inspected. The structural members are located and the mounting hardware is installed and tightened to the manufacturer's specifications. When all the hardware is installed, the panels themselves are installed and wired into the other components to form the complete photovoltaic system. Inverters and charge controllers must be mounted, programmed, and verified as operational. Finally, all wiring connections are made and the system is labeled per code requirements. A final system check is performed to ensure the proper operation of all components.

1. Some of the most important things to review when preparing for a solar panel installation job are the building permits.

    a. True
    b. False

2. The low end of a rafter will be physically closer to which of the following parts of a roof?

    a. Span
    b. Soffit
    c. Riser
    d. Ridge

3. When the solar panel equipment arrives on site, which items need to be stored close to the work area because they will be installed first?

    a. Mounting hardware
    b. Solar panels
    c. Inverters
    d. Batteries

4. Ladders used to access a roof must extend at least how far above the surface of the roof?

    a. 18"
    b. 24"
    c. 36"
    d. 42"

5. For electrical safety, how many feet of clearance space are generally acceptable for most PV components?

    a. 2
    b. 3
    c. 4
    d. 6

6. When preparing to install a Quick Mount PV® anchor from UNIRAC®, what size pilot hole should be drilled to get the anchor bolt or screw started?

    a. ¹⁄₁₆"
    b. ¼"
    c. ⁵⁄₁₆"
    d. ⅜"

7. The use of PV wire and cable as conductors of PV electricity is addressed in _____.

    a. *NEC Section 110.26*
    b. *NEC Section 690.35*
    c. *NEC Section 705.12*
    d. *NEC Section 708.52*

8. When marking rafters, remember that their edge is actually less than _____.

    a. 1" thick
    b. 1½" thick
    c. 2" thick
    d. 2½" thick

9. When splicing beams using CLICKSYS™ beams and attachments from UNIRAC®, the nuts and bolts should be tightened to _____.

    a. 7.5 foot-pounds
    b. 10 foot-pounds
    c. 25 foot-pounds
    d. 75 foot-pounds

10. UNIRAC® beams should be installed in spans not to exceed _____.

    a. 12'
    b. 24'
    c. 36'
    d. 48'

11. The output of a combiner is normally fed into a(n) _____.

    a. DC disconnect switch
    b. AC disconnect switch
    c. charge controller
    d. inverter

12. As an FLA battery is charging, its temperature should never exceed _____.

    a. 77°F
    b. 93.5°F
    c. 110°F
    d. 115°F

13. A Square D disconnect with a switch nameplate of 30A will have a DC rating of _____.

    a. 10A per pole
    b. 20A per pole
    c. 60A per pole
    d. 100A per pole

14. When terminating the PV+ wire to the PV+ terminal inside an Outback FLEXmax 60 charge controller, the PV+ lug should be tightened to _____.

    a. 17 inch-pounds
    b. 20 inch-pounds
    c. 25 inch-pounds
    d. 35 inch-pounds

15. When wiring in a Sunny Boy 3000US inverter, it must be tied to the _____.

    a. combiner
    b. charge controller
    c. AC circuit breaker
    d. AC ground for the utility

16. When installing the PV panels, the bolts and nuts holding the panels in place must be torqued to the manufacturer's specifications, which are typically _____.

    a. 25 foot-pounds
    b. 50 foot-pounds
    c. 75 foot-pounds
    d. 100 foot-pounds

17. Which section in the *NEC®* states that a sign must be placed at the service-entrance equipment indicating the type and location of on-site optional standby power sources?

    a. *NEC Section 110.26*
    b. *NEC Section 690.35*
    c. *NEC Section 702.7*
    d. *NEC Section 720.3*

18. When PV equipment is set up under standard test conditions and placed under no load, the short-circuit currents should remain within what percentage of each other as long as the irradiance on the panels stays the same?

    a. 2
    b. 5
    c. 8
    d. 10

19. Under standard test conditions, the temperature on the surface of the panels should be _____.

    a. 112°F
    b. 131°F
    c. 200°F
    d. 212°F

20. When all the individual components losses are calculated for a 2 kW PV system using no storage (batteries), the system's final output should be roughly what percentage of the sum of the rated power of the panels?

    a. 60 to 65 percent
    b. 70 to 80 percent
    c. 85 to 88 percent
    d. 88 to 92 percent

Marcel Veronneau

CEO IMTI Waterbury, CT

"When you can look at what you've done and know it's good, know you did it well and it will do what it's supposed to because you made it that way—that's craftsmanship."

How did you get started in the construction industry?
I started out as an electrical apprentice. It was a four-year program with an industrial background.

Who inspired you to enter the industry?
The people around me when I was growing up. They were already in the industry and seemed to be doing well. I was young and restless, and working with my hands appealed to me. There was a lot of variety in what they did, and I figured there was room to grow in different directions if I wanted to.

What do you enjoy most about your job?
I still enjoy variety, but as I look back, it's been a satisfying career. I worked hard and was rewarded for it. I always enjoyed teaching the new kids too, showing them the right way. The safe way. The opportunities came and I got into training. That's satisfying—watching the people I helped train move into their own careers.

Do you think training and education are important in construction?
Our industry is changing rapidly with technological advances and the emphasis on new forms of energy. The equipment I started out with is very different from the equipment of today. You have to take advantage of every opportunity to learn. If you don't keep up, you'll be standing on the sidelines watching the world go by.

How important are NCCER credentials to people just starting out?
Very. Our society is more mobile now. Those credentials are recognized and respected all over. They'll let you move anywhere in the country and show that you know your field and can do the job.

How has training or construction impacted your life and career?
It's let me raise a family, have a home, and generally a good life. I've been able to advance steadily and had a good time doing it.

Would you suggest construction as a career to others?
Yes. It's satisfying to build or have a part in something real that you can point to.

How do you define craftsmanship?
When you can look at what you've done and know it's good, know you did it well and it will do what it's supposed to because you made it that way—that's craftsmanship.

JOB SAFETY ANALYSIS

JOB SAFETY ANALYSIS

TITLE OF JOB OR TASK

TASK	START	END	HAZARDS	CONTROLS
1.				
2.				
3.				
4.				
5.				
6.				
7.				
8.				

Required Training:

Required Personal Protective Equipment (PPE):

Weekly Vehicle Check List: _____ Tire Pressure _____ Transmission Fluid
_____ Oil _____ Lights
_____ Air Filter _____ Wkly Mileage

Job Name:
Job Number:
Supervisor:
Date:

PRINT NAME	SIGN NAME	TOTAL HOURS

Names of Employees:

104A01.EPS

Additional Resources

This module presents thorough resources for task training. The following resource material is suggested for further study.

AMtec Solar (combiners) website: www.amtecsolar.com.

Florida Solar Energy Center website: www.fsec.ucf.edu.

Solar panel terms: www.osha.gov.

Surrette/Rolls Battery website: www.surrette.com.

Solar Source Institute website: www.solarsource.net.

University of Oregon's Solar Radiation Monitoring Laboratory website: http://solardat.uoregon.edu.

Figure Credits

Safety Hoist Company, Figure 2 and SA01

Topaz Publications, Inc., Figures 7, 17, 36, 58, and 60

Fall protection materials provided courtesy of Miller Fall Protection, Franklin, PA, Figure 10

Concorde Battery Corporation, Figure 11

Reimann & Georger Corporation, Figure 12

Surrette Battery Co. Ltd., Figure 13

Sharp USA, Figures 15 and 28

U.S. Department of Energy, Office of Energy Efficiency and Renewable Energy, Figure 16

AMtec Solar, Figures 18 and 37

www.solarabc.org/permitting Solar America Board for Codes and Standards, Figure 20

Canadian Solar Inc., Figure 21 (photo)

Unirac, Inc., Figures 22, 24, 26, 29, 30, and 32

Quick Mount PV, Figures 23 and 34

Atlantis Energy Systems Inc., Figure 25 (integrally-mounted panels)

John A. Bloom, Figure 25 (standoff-mounted panels)

ProtekPark Solar, Figure 25 (ground-mounted panels as carport covers)

Sun Banks Solar, Figures 25 (pole-mounted panels) and 35 (nameplate close-up)

Mike Powers, Figures 31, 35 (underside of panel), 38, and SA02

Square D/Schneider Electric, Figure 39 and Table 3

Outback Power Systems, Figure 41

SMA Solar Technology AG, Figures 44–53 and 55

Tyco Electronics Energy Division, Figure 59

L. J. LeBlanc, Appendix

NCCER CURRICULA — USER UPDATE

NCCER makes every effort to keep its textbooks up-to-date and free of technical errors. We appreciate your help in this process. If you find an error, a typographical mistake, or an inaccuracy in NCCER's curricula, please fill out this form (or a photocopy), or complete the online form at **www.nccer.org/olf**. Be sure to include the exact module ID number, page number, a detailed description, and your recommended correction. Your input will be brought to the attention of the Authoring Team. Thank you for your assistance.

Instructors – If you have an idea for improving this textbook, or have found that additional materials were necessary to teach this module effectively, please let us know so that we may present your suggestions to the Authoring Team.

NCCER Product Development and Revision

13614 Progress Blvd., Alachua, FL 32615

Email: curriculum@nccer.org
Online: www.nccer.org/olf

❏ Trainee Guide   ❏ Lesson Plans   ❏ Exam   ❏ PowerPoints   Other _____

Craft / Level: _____   Copyright Date: _____

Module ID Number / Title: _____

Section Number(s): _____

Description: _____

_____

_____

_____

Recommended Correction: _____

_____

_____

_____

Your Name: _____

Address: _____

_____

Email: _____   Phone: _____

Maintenance
and Troubleshooting

57105-11

Trainees with successful module completions may be eligible for credentialing through NCCER's National Registry. To learn more, go to www.nccer.org or contact us at **1.888.622.3720.** Our website has information on the latest product releases and training, as well as online versions of our *Cornerstone* magazine and Pearson's product catalog.

Your feedback is welcome. You may email your comments to **curriculum@nccer.org**, send general comments and inquiries to **info@nccer.org**, or use the User Update form at the back of this module.

 V.2 11/13

Objectives

When you have completed this module, you will be able to do the following:

1. Identify the tools and equipment required for maintaining and troubleshooting PV systems.
2. Measure system performance and compare to expected performance.
3. Perform system maintenance as recommended by the PV equipment manufacturer.
4. Perform diagnostic procedures, interpret the results, and implement corrective measures on a malfunctioning system.
5. Verify system functionality, including startup, shutdown, normal operation, and emergency/ bypass operation.
6. Compile and maintain records of system operation, performance, and maintenance.

Performance Tasks

Under the supervision of the instructor, you should be able to do the following:

1. Demonstrate typical maintenance procedures on an installed PV system and document the results.
2. Troubleshoot a malfunctioning system and document the results.

Trade Terms

Specific gravity
Refract

Prerequisites

Before you begin this module, it is recommended that you successfully complete *Core Curriculum* and *Solar Photovoltaic Systems Installer*, Modules 57101-11 and 57102-11. It is also suggested that you shall have successfully completed the following modules from the Electrical curriculum: *Electrical Level One*, Modules 26101 through 26111; *Electrical Level Two*, Modules 26201, 26205, 26206, and 26208 through 26211; *Electrical Level Three*, Modules 26301 and 26302; and *Electrical Level Four*, Modules 26403 and 26413.

Note: *NFPA 70®*, *National Electrical Code®*, and *NEC®* are registered trademarks of the National Fire Protection Association, Inc., Quincy, MA 02269. *All National Electrical Code®* and *NEC®* references in this module refer to the 2011 edition of the *National Electrical Code®*.

Contents

Topics to be presented in this module include:

Figures and Tables

1.0.0 INTRODUCTION

After photovoltaic (PV) system components are installed and operational, they require routine preventive maintenance. Environmental conditions affect the way the equipment works. Excessive heat or cold and rain or snow changes the way the equipment generates power. Lightning and hail can damage PV equipment. In addition, components wear out over time. PV components must be inspected regularly to ensure proper system operation and to perform scheduled maintenance.

This module explains how to make operational checks and maintain PV equipment to help it last longer and operate more efficiently. The results of operational checks are compared with baseline data to determine the efficiency of the equipment and spot potential problems before they result in equipment failure.

Regardless of whether a PV system has one panel or a hundred, maintenance and troubleshooting are essentially the same. The output of the inverter should be equal to the output from the array panels, minus any losses created by the individual PV components. *Figure 1* shows a simple diagram of the items to be maintained.

PV systems produce different levels of DC power (voltage and current), depending upon how the equipment is designed and wired. Although the voltage levels may be under 50VDC on most residential systems, there is always the possibility of higher voltages and currents. All PV equipment must be properly grounded and Cat III/IV meters must be used. Only qualified individuals may perform PV system maintenance and troubleshooting. Per *NEC Article 100*, a qualified person is one who has the skills and knowledge related to the construction and operation of the electrical equipment and installation, and has received safety training to recognize and avoid the hazards involved. Capable system owners and operators may be trained to do limited maintenance, which includes making visual inspections and operational checks. Anything more than that should be performed by persons qualified to work on PV systems.

WARNING!

Safety is the most important consideration when performing PV system maintenance and troubleshooting. Use fall protection when working at heights of six feet or more. Follow lockout/tagout procedures when working with electrical equipment. Always wear appropriate personal protective equipment.

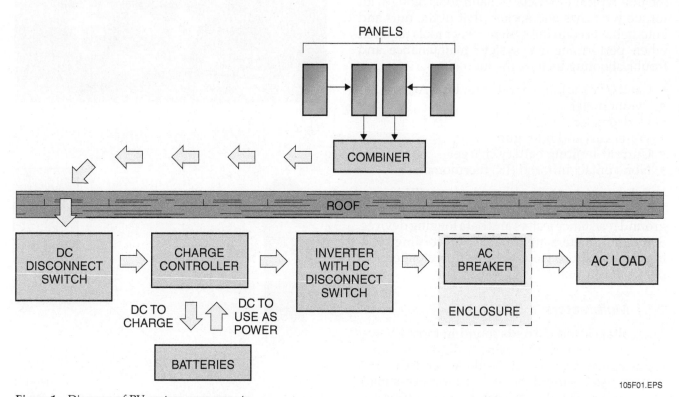

105F01.EPS

Figure 1   Diagram of PV system components.

2.0.0 PREVENTIVE MAINTENANCE

PV panels and other system components are installed outside and exposed to all weather conditions. They must be able to withstand various environmental stresses, including extreme temperatures and wind. Wiring from the panels may be run through conduit before being tied into the combiner. From the combiner on down, the wiring should be protected by conduit that connects each of the PV system components (disconnect switches, charge controller, and inverter). The charge controller, batteries, inverter, and any disconnects should be sheltered from the weather.

> **WARNING!**
>
> Any work inside the enclosure of a PV system component is potentially dangerous. Operational checks must be made with power applied, which means that electrical shock inside the enclosures is always a possibility. Follow safe work practices and wear the appropriate PPE when working around energized equipment. Use insulated tools.

2.1.0 Tools and Test Equipment

The tools needed for PV maintenance include appropriate PPE and ladders or lifts for gaining access to the panels. The hand tools required include typical electrician's hand tools along with torque wrenches and sockets that fit the nuts and bolts being used in the system. Other tools required when performing PV system maintenance and troubleshooting include the following:

- Cat III/IV multimeter rated for both AC and DC
- Pyranometer
- Hydrometer
- Water cart and filler gun
- Current-limiting battery charger
- Non-contact infrared (IR) thermometer

Advanced testing and analysis of PV equipment may require special equipment such as ground resistance testers and data logging devices. Some multimeters are capable of recording readings so that those values can be downloaded to a personal computer.

2.1.1 Multimeters

The voltages and currents found in most PV systems can be measured using a digital multimeter designed for use with both AC and DC. The meters used should be true root-mean-square (rms) types capable of measuring continuity,

resistance, and up to at least 600 VAC/VDC. When DC currents over 10A must be measured, a separate ammeter or a multimeter capable of measuring the higher currents is needed. *Figure 2* shows a multimeter that can be used to measure currents up to 1,000A.

When low-voltage testing devices such as pyranometers produce output signals in the millivolt range, a multimeter with at least 4½-digit resolution may be needed to attain a desired level of accuracy. That level of accuracy may come in handy when checking solar panels (*Figure 3*). The black test lead is attached to the negative (–) wire from the solar panel. The red test lead is attached to the positive (+) wire from the solar panel. If the solar panel has been, and is still, completely covered, the voltage will be minimal.

Multimeters can also be used to check continuity of the cabling or wiring, the amount of resistance across a PV device, or the current (amperage) being pulled through a wire. Some

105F02.EPS

Figure 2 Multimeter.

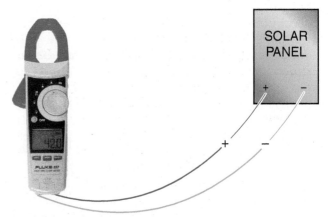

105F03.EPS

Figure 3 Multimeter checking voltage across a solar panel.

NCCER — *Contren® Learning Series* 57105-11

multimeters can measure the frequency of a signal when its jaw clamps surround a power cable. For most multimeters to read the frequency of a power line signal, the wire must be carrying at least 5A of current. Knowing the frequency of the output power of a PV system's inverter is important since that output must be the same as the grid, or the proper frequency of the power needed by conventional AC appliances. *Figure 4* shows different ways to use a typical multimeter.

2.1.2 Pyranometers

Pyranometers (*Figure 5*) are devices used to measure solar irradiance detected on a flat, two-dimensional surface. That means that they take in a 180-degree view and determine the intensity of the sunlight on the area. Some pyranometers are installed as fixed devices. Others are mobile test equipment items often used by maintenance personnel.

If a handheld pyranometer (*Figure 6*) is being used, position it so that its sensor area is in the same orientation as the PV panels. After its readout has stabilized, record the irradiance value. Remember that solar irradiance is measured in watts per square meter.

After the irradiance value has been determined, the next step is to determine the panel area (not including the frame) in square meters. Multiply the area of the panel or the array by the irradiance reading to find the power (in watts) that the panel or array should be producing. Consult the panel manufacturer's data to determine the conversion efficiency of the panels. This data may also list the amount of active area on a panel. Multiplying the panel nameplate power times the efficiency value results in an estimated amount of power that should be produced by the panels. When a multimeter is used to measure the power from a panel or an array tied into a combiner, the multimeter readout should be very close to the calculated value.

Conduct operational checks on clear days when the sun is located directly overhead. The panels should be as clean as possible. As conditions change throughout the day and over the seasons, the output power from the panel or array also changes. If the panels are obstructed by leaves, dust, pollen, or bird droppings, the output power will be less than when the panels are clean. When the panels are new and clean, tests are run to provide a baseline of data for future preventive maintenance or troubleshooting checks. As the panels age, their power output will gradually diminish. The life expectancy of well-maintained PV panels can be 20 years or more.

2.1.3 Hydrometers

PV systems may or may not use storage batteries for backup power. But if batteries are used, they must also be maintained. Other than possibly needing their terminals cleaned once in a while, sealed deep-cycle lead-acid batteries are maintenance free. Sealed gel-cell batteries are almost maintenance free, but do need a little maintenance to keep them at their peak. Flooded lead acid (FLA) batteries, also known as wet cell batteries, require the most maintenance.

The flat lead plates inside the cells of these batteries must be immersed in electrolyte for them to retain and discharge voltage. The electrolyte is a mixture of water and acid. That mixture of acid and water gives the electrolyte its specific gravity value. As a battery charges and discharges, some of the water evaporates, changing the specific gravity of the cells. Water must be added periodically to maintain the specific gravity of the batteries at the desired level.

The specific gravity of the FLA batteries used in PV systems is measured by a hydrometer. There are two basic types of hydrometers: the Archimedes type and the refractive index type. The Archimedes type is based on the principle that the buoyant force on a submerged object is equal to the weight of the liquid displaced by the object. With this type of hydrometer, a sample of electrolyte, usually a few ounces, is drawn from each cell into a float chamber. The higher density and specific gravity of the electrolyte, the higher the float rises, and specific gravity is measured using calibrated graduations on the float or chamber. Due to the effects of temperature on specific gravity, adjustments to the reading must be made if the electrolyte is significantly warmer or cooler than room temperature. Modern hydrometers are designed to compensate for the temperature surrounding the sample and the hydrometer being used. *Figure 7* shows an Archimedes-type digital hydrometer that can store over 1,000 individual readings. Those readings can later be downloaded onto a computer programmed with associated software. Such records are good maintenance history documents.

The second type of hydrometer is called a refractive index hydrometer or a refractometer. It uses a prism and the principle that light refracts at greater angles through solutions with higher specific gravity. A small drop of electrolyte is all that is needed for the measurements. Manual refractive index hydrometers, also known as refractometers, have a scale inside the viewing area that allows the operator to see how the sample shows up on the scale. *Figure 8* shows a typical refractometer.

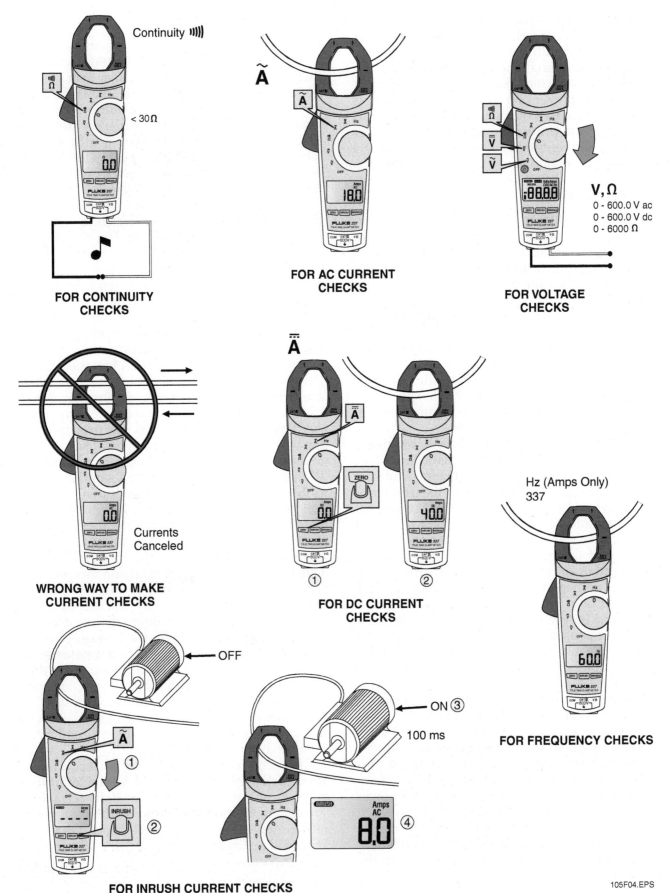

Figure 4   Ways to use a typical multimeter.

PYRANOMETER MOUNTED ON TEST STAND IN FIELD

105F05.EPS

Figure 5   Pyranometers.

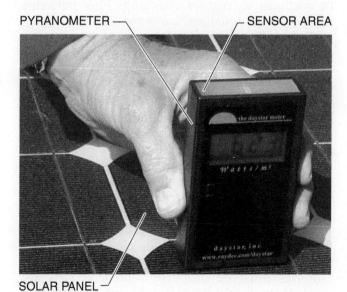

PYRANOMETER — SENSOR AREA

SOLAR PANEL SURFACE

105F06.EPS

Figure 6   A handheld pyranometer positioned on a solar panel.

READOUT

CONTROLS

SAMPLE TUBE MOUNTING LOCATION

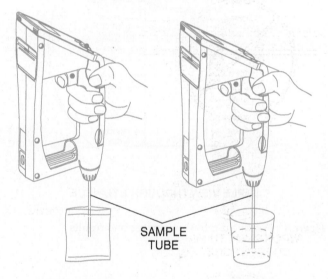

SAMPLE TUBE

SAMPLE TUBE IN TEST SOLUTION

TAKING IN A SAMPLE

105F07.EPS

Figure 7   Digital hydrometer.

The drawback of manual refractometers is that they rely on human interpretation of the results. Digital refractometers are also available.

2.1.4 Water Carts and Filler Guns

As an FLA battery is repeatedly charged and discharged over time, some of the water in the electrolyte is lost. More distilled water must be added. A convenient and safe way to add water to FLA batteries is with a filler gun specifically made for batteries (*Figure 9*). That gun is attached to the

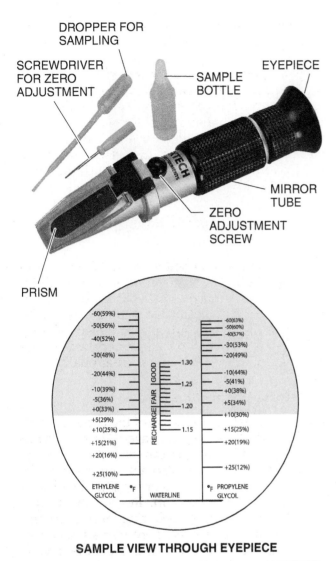

DROPPER FOR SAMPLING

SCREWDRIVER FOR ZERO ADJUSTMENT

SAMPLE BOTTLE

EYEPIECE

MIRROR TUBE

ZERO ADJUSTMENT SCREW

PRISM

SAMPLE VIEW THROUGH EYEPIECE

105F08.EPS

Figure 8 Handheld refractometer use and results.

end of the hose from a water cart. Some carts have built in pumps powered by an onboard battery system. Others are powered from an external 110 VAC power source. Using a filler gun allows the water to be added slowly to the battery cells.

> **WARNING!**
>
> Be careful when adding water to the battery cells because the electrolyte can splash back out. Always wear the appropriate PPE for handling acids and make sure that a spill kit is available. Acids can cause burns and destroy clothing.

The components inside the batteries essentially become stale with use over time. Discharging FLA batteries deeply and leaving them discharged for extended periods of time will ruin them. Allowing the plates to dry out will result in chemical changes. Batteries that have been deeply discharged and then only partially recharged on a regular basis may fail in less than a year. If batteries are deeply discharged and then quickly recharged, the specific gravity reading will be lower than it should be because the electrolyte near the top of the battery has not had time to properly remix with the charged electrolyte deeper in the battery. Battery manufacturers recommend that the electrolyte be checked at least every quarter. Always use distilled water to replenish the batteries.

2.1.5 Current-Limiting Battery Chargers

In addition to regular charging, FLA batteries must also be equalized on a regular basis (normally once per month). Equalizing a battery is a process in which the cells are overcharged at a steady controlled rate. The overcharging process heats the electrolyte and makes it bubble. The bubbling electrolyte removes any lead sulfate buildup from the plates inside the battery, which improves the charge/discharge rate. The charge controller in a PV system normally equalizes the solar storage batteries; however, some PV systems require the use of an external current-limiting battery charger. These chargers are similar to those used on automotive batteries.

During the charging and overcharging process, the bubbling electrolyte gives off vapors. Those vapors are toxic and can cause an explosion. Batteries must be properly vented. If fans are used, they should be mounted outside blowing in and vented out at the top.

Other types of solar batteries may also require equalization, but it is performed less frequently

105F09.EPS

Figure 9 Filler gun.

and only as directed by the battery manufacturer. Always refer to the manufacturer's maintenance instructions for the batteries in use.

2.1.6 Non-Contact IR Thermometers

Batteries give off additional heat as they charge, or as they are being equalized. Some PV systems have built-in sensors to monitor the temperature of the batteries during these processes. For maintenance purposes, use non-contact infrared (IR) thermometers to check the temperature. Another name for these devices is non-contact pyrometers. *Figure 10* shows a non-contact IR thermometer with a laser beam.

A non-contact IR thermometer can be used for a variety of temperature measurements required during PV installation and maintenance. These instruments can be used to measure the temperature of panels and arrays, electrical equipment, connections and terminations, electrolyte, and other fluids and surfaces. A red laser beam is used to aim the device. Thermal radiation sensors within the instrument approximate the surface temperatures within a specified distance and field of view. These thermometers are reasonably accurate within their range of operation and have an advantage in that they do not require touching the equipment. More detailed thermal investigations of entire PV arrays,

electrical systems and equipment, and building envelopes may be conducted using a thermal imaging camera (*Figure 11*).

Power and energy measurements are an important part of system installation, operation, and maintenance. Most inverters now include some form of onboard monitoring of power, and record daily and total energy production. In other cases, external watt-hour meters are used, which can be electronic or electromechanical, similar to a standard utility service meter.

2.2.0 Cleaning and Inspecting PV Equipment

All equipment works better when it is clean and well maintained. PV installation and maintenance personnel may be called back to perform periodic cleaning and inspections of installed PV equipment. Such preventive maintenance actions are often carried out on a quarterly basis, but should be performed at least semi-annually.

2.2.1 Visual Inspections

Start with a visual inspection of the panels. Check for cracked glass and damaged frames. Wind forces and falling debris can distort or damage the panels. Wind can also vibrate the panels, eventually loosening the clamping hardware and even the anchoring system. Review the torque requirements for the anchoring and clamping hardware, and then make sure that all fasteners are properly tightened. Check to see if the threads have been marked. If they were marked, and the nuts or bolts have not moved (the marks are still aligned), they should be tight. If they must be retorqued, remember to loosen the nut or bolt

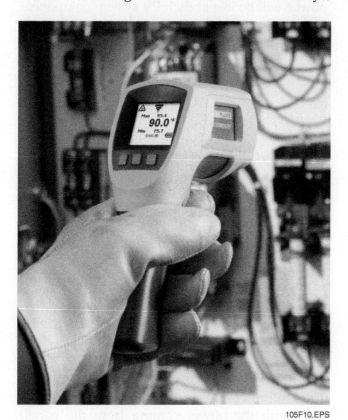

105F10.EPS

Figure 10   Non-contact IR thermometer with laser.

THERMAL HOT POINT

105F11.EPS

Figure 11   A thermal imaging camera being used to check breakers and wiring.

 Maintenance and Troubleshooting

slightly before retorquing it. While checking the anchoring devices, make sure that the weatherproofing materials around any anchors penetrating a roof are not dried out or damaged. Add more weatherproofing material as needed.

Check all cables coming into and out of the panel's terminal or junction box. Make sure that all cables are securely fastened to the framework supporting the panels. Verify that any combiners used are solidly anchored. Open combiner covers and make sure that the connections and wiring are securely fastened and undamaged. Check for any signs of arcing.

> **WARNING!**
> Unless the disconnect switch feeding into a PV system enclosure is opened, the wiring and terminals inside the enclosure are energized. Follow safe work practices and wear all appropriate PPE for working on or near exposed circuits.

Check all conduit for corrosion and tightness. Make sure that it is properly anchored. Check the labeling on all PV components to make sure that it is installed and is still legible. Check the inside and outside of all disconnect switches, charge controllers, and inverters, and verify that all equipment is clean and undamaged. Also, make sure that all wiring is undamaged and is properly installed. Check the batteries and their terminals, cable connectors, and cables for damage. Make sure that the batteries are not leaking from overflows. Verify that the venting system used with the batteries is functional and unobstructed.

2.2.2 Cleaning

The output from a solar array depends on the amount of sunlight received by the panels. Dirty panels cannot operate properly. Before attempting to evaluate the performance of a PV system, ensure that the panels are clean. Refer to the panel manufacturer's instructions for recommended cleaning materials and procedures. In most cases, a mixture of water and dishwashing detergent can be used to loosen dirt and bird droppings. The soapy mixture is sprayed onto the panels, allowed to soak, and then rinsed off with clean water.

> **CAUTION**
> Do not use a pressure washer to clean PV panels. This equipment can damage the panels, weathersealing, and wiring.

The batteries may also need to have their exteriors cleaned. Exercise caution and wear the appropriate PPE. Remove any accumulated dust and check for loose connections and signs of leakage.

> **WARNING!**
> When cleaning batteries, wear the appropriate PPE for handling acids and make sure that a spill kit is available. Acids can cause burns and destroy clothing. Some batteries may have acid residue on their exteriors that will eat through most cloths. If cloths are used to wipe down the batteries, dispose of them in a contaminated waste container.

2.3.0 Evaluating Performance of PV Systems

Each PV system should have some form of documentation indicating the level of performance expected at each PV component input and output. The system owner should have those records on file. These records are the standards against which the system's current performance should be judged. If maintenance records from previous inspections are available, review them and look for performance trends. Make notes as to what each of the individual PV components in a system should be producing. Remember that some power will be lost at each PV system component. Make sure to conduct the evaluations when the environmental conditions around the PV system are as close as possible to Standard Test Conditions (STC). To keep from loading the PV system, disconnect any batteries from the charge controller and open the AC breaker on the output of the inverter.

2.3.1 Overall PV System Check

After reviewing data from past records, proceed to test and record the values from the overall PV system and each individual PV component. Before checking the individual PV components, record and compare the output of the inverter. For a utility-interactive system with no storage, the inverter output should be within 70 to 80 percent of the overall output from the array when it was new. If the overall system output is less than 70 percent, one or more of the PV system components may be causing the loss.

2.3.2 Panel Checks

To check the individual PV system components, start with the power coming from the panels. Use the appropriate test equipment to obtain the voltage and current values from the individual panels. Access to those panel outputs is available at the input terminals of the combiner. If a combiner has a circuit breaker for each incoming leg, open the breaker so that no backfeed can influence the reading taken from a given panel. If the combiner is equipped with fuses instead of circuit breakers, the fuse used with a given input could be removed to accomplish the same result. Record the voltage and current from each panel feeding the combiner. Compare these values with the expected values. Determine if any of the panels must be replaced. Replace the fuses or close the circuit breakers before checking the output of each combiner.

2.3.3 Combiner Checks

With the inputs from the panels known, calculate the total input to the combiner. Consult the system records to determine the expected loss through the combiner under ideal conditions. Use that efficiency factor to determine the expected output for the combiner. Make the needed measurements and record the findings. Compare the actual values to the calculated values and determine if the panels and the combiner are operating as planned. If the output of the combiner is not within the expected range, determine if the combiner needs to be replaced. Verify that each combiner is properly labeled.

2.3.4 Wiring and Raceway Checks

The visual inspection should reveal any loose or damaged exposed wiring. Without disconnecting the wiring and removing it from the raceway or conduit sections, there is not much that can be done with the wiring and raceway connecting the PV components. If it appears that a section of wiring is dropping too much power, specific tests can be run on the wiring.

2.3.5 Disconnect Switch Checks

The DC disconnect switches located at the output of the combiners and at the input of the inverters are basically go or no-go devices. Some may have fuses, but most do not. A check of these switches should include a visual inspection of the wiring coming into and going out of the switch enclosures. Their terminals should appear to be tight and undamaged. If the switch contacts can be seen, they should appear to be clean without traces of smoke or arc strikes. These switches are rarely turned off and on, but they do need to be checked to see that their mechanical components are functional. Check the user manual to see if the

mechanical parts that move require any kind of lubricant. Lubricate according to the user manual. With incoming power applied to a disconnect switch, turn the switch off and use a multimeter to verify that no power is crossing the switch contacts. Verify that the switch can be locked out. Close the switch contacts and verify that the switch is transferring power from the input side to the output side. Some manufacturers recommend that switch contacts be opened and closed ten times (exercised) on an annual basis to keep the contacts lubricated. Verify that all disconnect switches are properly labeled.

2.3.6 Charge Controller and Inverter Checks

The charge controller and the inverter are the only two PV components that are programmable. Refer to the user manual for each device, along with any site-specific programming documentation, to verify that the charge controller and inverter are properly programmed and functioning as designed.

Review the user manual and any site-specific documentation to determine how much power should be lost across a charge controller or inverter. Remember that the charge controller should initially be checked without any connection to the batteries. In most cases, PV charge controllers and inverters should both have an output that is about 95 percent of the input. Measure the power coming into one of these devices, subtract the expected loss factor, and then see how close the output comes to the calculated value. Record the findings and compare them against the system records to see if the devices are still operating as designed.

Disconnect the inverter from any AC load. Apply power to it, either directly from the array or from the charge controller without batteries connected. Check and record the DC voltage and current applied to the inverter. Check and record the AC voltage and current produced by the inverter. Connect the output of the inverter to an AC load, and then repeat the inverter input and output checks. Verify that the inverter is producing the desired AC power. If not, adjust or replace it as needed.

2.3.7 Storage Battery Checks

The amount of power that actually reaches the storage batteries is always a bit less than the output of the array. The system components and wiring drop some of the power as it is processed from the array to the batteries or to the inverter. The weather also affects the amount of power provided to the batteries.

The type of charge controller also makes a difference. Unless the charge controller is a maximum power point tracking (MPPT) charge controller, the amount of power applied to the batteries is about 80 percent of the output of the array. If an MPPT charge controller is used, the batteries should receive about 95 percent of the power from the array. MPPT charge controllers take any excess voltage created by an array and create a higher charge current than normal charge controllers. When checking batteries, determine the type of charge controller being used and adjust the values accordingly.

Disconnect the output of the inverter from any AC load. Change the charge controller so that it supplies the inverter with only power from the batteries. Measure and record the amount of battery power reaching the inverter. Measure and record the amount of AC power from the inverter as it is receiving battery power. Close the AC breaker and connect the AC load to the output of the inverter. Measure and record the amount of inverter AC power under load. Compare those numbers against the design values to see if the batteries and the inverter are properly supporting the AC load.

2.3.8 Grounding and Lightning Arrestor Checks

All PV systems must have good grounding. They must also have good operational lightning arrestors. When inspecting PV equipment, verify that all grounding devices are securely attached and undamaged. The same goes for the lightning arrestor devices used on the PV equipment. Remember that the individual solar panels must be tied into the overall grounding system. Verify that all grounding clamps are securely fastened without any corrosion.

3.0.0 MANUFACTURER-RECOMMENDED MAINTENANCE ACTIVITIES

For the purpose of maintenance review, this section focuses on four panels, one combiner with a DC disconnect switch, one charge controller, two FLA batteries, and one inverter with a DC switch on its input and an AC breaker on its output. Not every work center or technical school will have the same devices addressed in this module, but the principles discussed still apply. The PV components that require regular inspection and maintenance include the following:

- Solar panels
- Mounting hardware
- Combiners
- Disconnect switches
- Charge controller
- Batteries
- Inverter

3.1.0 Recommended Panel and Mounting Hardware Maintenance

Ideally, PV systems should be inspected every three months at the change of seasons. Review the system's history records before going onto a PV system installation site. When inspecting the array, start with the attachments holding the mounting hardware in place.

3.1.1 Mounting Hardware Checks

Determine the type of panels and mounting hardware being used. Check the manufacturer's specifications for the torquing levels of the mounting hardware. Most torque levels are less than 100 inch-pounds. Make sure that the correct tools are on hand for the inspection. If the panels are mounted close to a roof, take along a strong flashlight for better inspections in dark areas. Also, include the tools or solutions needed for cleaning and repairing any corroded areas.

If the array is mounted on a roof, go under the roof (possibly inside an attic) to verify that nothing inside is damaged or torn loose. While checking the inside or underside, also check for leaks that may have been caused by the mounting hardware or conduit penetrations through the roofing materials. *Figure 12* shows conduit that has been run through a roof and into an attic.

On the outside of a roof or a stand-alone rack, verify that all mounting beams, standoffs (if used), clamps, nuts, and bolts are undamaged, securely fastened, and not corroding. Galvanic corrosion is usually caused by dissimilar metals contacting one another. Orange/brown streaks that appear to run down a surface are a good indication of galvanic corrosion. If allowed to continue, corrosion will weaken the support system. Mounting hardware is designed and built to be lightweight, but strong. If corrosion weakens the lightweight hardware, one or more of the panels could break loose from the mounting. If corrosion is found, clean it and make any needed repairs. *Figure 13* shows examples of mounting hardware.

While inspecting the mounting hardware, check under the panels and remove any collected debris such as leaves or trash. Dry trash and leaves can be a fire hazard. Also, check for and remove

Figure 12 Conduit inside an attic.

STANDOFF WITH WEATHER SEAL

BEAM

FRAME OF SOLAR PANEL

BOLT AND NUT FOR CLAMP

BEAM

105F13.EPS

Figure 13 Mounting hardware.

any vegetation growth around and under the panels. Verify that all weathersealing attachments and materials are in good condition. *Figure 14* shows an example of weathersealing attachments used to anchor mounting hardware for PV arrays. If flashings are used, make sure that they are still sealing the roof.

Make sure that there is no protruding structural members, conduit, or other installed equipment that could trip or snag anyone moving near the panels. If unable to eliminate protruding equipment, mark it so that it is highly visible. *Figure 15* shows conduit and a combiner box mounted near an array installed on a flat roof. Notice the black and yellow tape covering conduit near pathways.

3.1.2 Mounting Hardware Grounding

Solar panels, and the hardware on which they are mounted, must be grounded and bonded together. *Figure 16* shows a simple diagram of the grounding on a grid-tied PV system. Other PV systems may have different components, but the idea is the same. Review the *NEC®* and any local electrical codes to ensure that the PV grounding guidelines are being applied to the system being inspected.

When checking the PV system grounding, make sure that the wiring is not broken, and that

STANDOFF MOUNTING HARDWARE

FLASHING

105F14.EPS

Figure 14 Weathersealing attachments under arrays.

it is securely attached to the mounting beams and panels. Repair as needed. *Figure 17* shows a close-up view of a properly installed ground wire connector.

Figure 15 PV conduit on roof.

Also, make sure that all jumpers and bonding devices used between the beams and panels are in place and securely fastened. If a panel must be removed for repair, the continuity of the grounding and bonding between the mounting hardware and the other panels must be maintained. *Figure 18* shows a WEEB bonding strap installed between two beams.

3.1.3 Solar Panel Checks

Solar panels, like anything else exposed to the elements, can be damaged by flying or falling objects or extreme temperature changes. Extreme temperature changes can cause metal frames to expand and contract. Such movements may

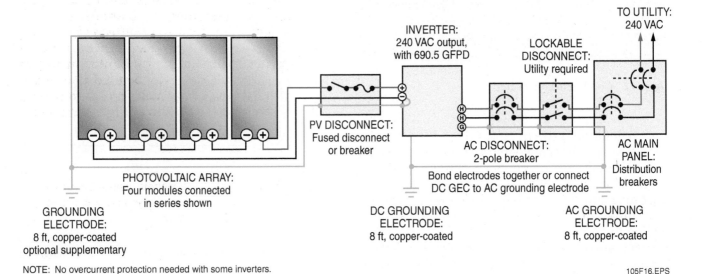

NOTE: No overcurrent protection needed with some inverters.

Figure 16 PV system grounding.

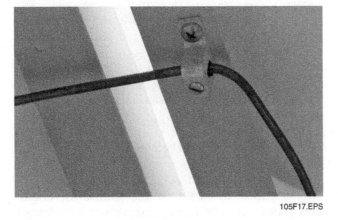

Figure 17 Ground wire connector.

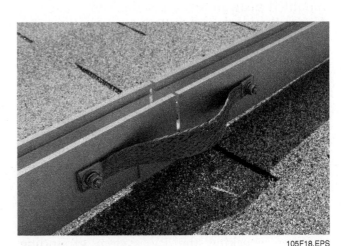

Figure 18 Beam bonding.

warp either the mounting hardware supporting a panel, or the framing around a panel itself. Falling branches, wind-blown debris, and hail can damage solar panels. If panels are mounted close to ground level, mowing machines could throw rocks into an array and damage panels. The surface area of a panel may have been cracked by a tool during installation and gone unnoticed then, but it may start to leak later.

If the laminate or glass protecting the circuitry inside a solar panel is cracked, the panel will absorb moisture or water. Water and electrical circuitry do not mix. The panel circuitry will develop high leakage currents and become unsafe. It will eventually fail. If a panel is getting moisture inside it, the ground-fault protection circuit for the panel or array will trip frequently. Such reactions indicate potential problems with the panels, the overall array, or the grounding system itself.

Dirty panels or panels covered in snow will both suffer significant output losses. *Figure 19* shows snow-covered solar panels.

Air contaminated with salt spray or acid rain can damage the panels. *Figure 20* shows a panel damaged by salt spray.

Acid rain is a broad term often used for wet and dry deposits containing higher-than-normal amounts of nitric and sulfuric acids. Acid rain may be produced by natural sources, such as volcanoes, but is more likely the result of manmade sources, such as fossil-fuel emissions.

One of the most critical things to do when inspecting and evaluating the solar panels is to make sure that they are not shaded, and that they are getting the proper amount of sunlight. Remember that any kind of shading greatly reduces the performance of the panels. Trees grow over time. New buildings may have been built since the PV panels were installed. Use a

105F20.EPS

Figure 20   Salt damage.

Solar Pathfinder™ or similar device to repeat the initial site assessment evaluations. This will show where any new shade is affecting the panels. A handheld pyranometer can be used to evaluate the quality of the sunlight available at the panels. A multimeter can be used to evaluate the individual panels that make up an array with multiple panels feeding into a single combiner.

> **WARNING!**
>
> Remember that solar panels cannot be turned off. They are always producing DC voltage at some given current level.

Before making any voltage checks on the panels, make sure that they have been cleaned. Also, make sure that the wiring or cables from a panel's modules to its junction or terminal box are securely fastened to the mounting beams. Cable clips or ties may loosen or come completely undone over time. Repair or replace the clips or ties as needed. While checking the panel cables, also check them for nicks, cuts, abrasions, or any kind of damaged insulation.

Check the + and − extension wiring or cables from the individual terminal boxes to the combiner shared by the panels. Look for loose connections and signs of damage. Check the inside of the combiner box and make sure that all input wires from the panels, and the output wires from the combiner, are securely fastened to the terminals. Look for signs of arcing or other damage. Clean the combiner as needed, and then close and secure the cover.

105F19.EPS

Figure 19   Snow-covered solar panels.

3.2.0 Recommended Combiner, DC Disconnect Switch, and Conduit Maintenance

Conduit may or may not be used between the terminal boxes on the panels and the combiner, but it is used between the combiner and other devices downstream from the combiner. The DC disconnect switch is most likely installed close to the inverter and is probably connected with a short section of conduit. Conduit is used from the disconnect switch on down to the charge controller (when one is used) and inverter. Any conduit installed outdoors is exposed to all weather conditions. Conduit must be coupled, at both the individual components and between sections. Those couplings must be raintight. When exposed to direct sunlight for long periods, the temperature inside the conduit becomes extremely high. Conduit expands as the temperature increases, and contracts as it decreases. Such physical changes may cause the conduit to buckle, which can break it loose from its anchor points. Such movements can also pull the conduit away from the couplings to individual PV components (combiner, disconnect switch, charge controller, or inverter).

The wiring or cables inside the conduit can also be damaged over time. Insulation breaks down under extreme heat. If a PV system is experiencing electrical shorts, check for damaged wiring.

Combiners and disconnect switches are contained in weatherproof enclosures with seals. Check the seals for damage any time the enclosure is opened. Inside the combiners or DC disconnect switches, check for loose or damaged wiring and signs of arcing. Make sure that the enclosures are clean and properly labeled. Vented enclosures must be screened or otherwise protected against the intrusion of insects, snakes, and rodents. All of the PV panel connections must be labeled with the warning: Do Not Disconnect While Under Load.

3.3.0 Recommended Charge Controller Maintenance

Charge controllers are designed for indoor installation. They should not be exposed to rain or direct sunlight. Most have air vents that must be cleaned on a regular basis. *Figure 21* shows the location of the air vents on the side of a typical charge controller.

Before doing any kind of preventive maintenance inside a charge controller, open all DC disconnect switches feeding power into the charge controller. If there is a disconnect switch between the output of the charge controller and the batteries or inverter, open that switch to prevent

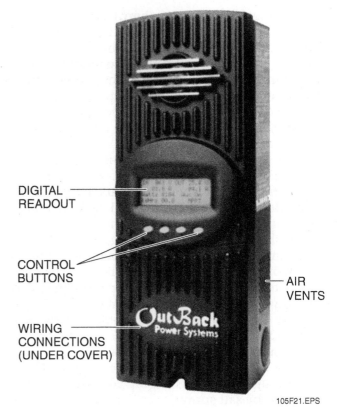

DIGITAL READOUT

CONTROL BUTTONS

AIR VENTS

WIRING CONNECTIONS (UNDER COVER)

105F21.EPS

Figure 21 Typical charge controller.

any backfeed from entering the charge controller while it is being cleaned and checked. Use a multimeter to verify that there is no voltage inside the charge controller. Review the torque specifications for all terminals inside the charge controller. Verify that all wiring into and out of the charge controller appears to be undamaged and properly installed. Check for any signs of arcing, corrosion, or loose connections. When all preventive maintenance work is completed, close the disconnect switches and return the charge controller to its normal operation.

Since each charge controller is different, refer to the user manual for the charge controller for specific operational checks that need to be done during scheduled maintenance activities. Primarily, checks need to be made to verify that the charge controller is charging the batteries properly, and properly applying DC power to the inverter it serves. If all these checks are good, continue on to the next component. If any problems are noted, troubleshoot and resolve them before proceeding.

3.4.0 Recommended Battery Maintenance

The batteries used in solar equipment are either sealed or vented. Sealed batteries are more expensive, but require less maintenance. Vented batteries are less expensive, but require more maintenance. Both types of batteries must be cleaned, and their terminals must be checked to ensure that the cables are securely attached. A multimeter is used to verify the DC voltage from the batteries.

<table>
<tr><td>WARNING!</td><td>Before working on any batteries, wear the appropriate PPE (goggles, gloves, boots, and a splash apron) for dealing with acids. Have a spill kit available. Also, make sure the work area is properly ventilated.</td></tr>
</table>

Before attempting to clean any vented batteries, make sure that their vent caps are properly installed and tightly fastened. Check for any signs of leakage around the battery. Leakage may indicate that the caps are not well sealed, or that the battery body may be leaking. Use a flashlight for closer inspections. If the batteries are not leaking and appear to be undamaged, use a mixture of water and baking soda (100g per liter) to clean them. Lightly wipe them down with the cleaning solution, and then rinse with clean water and dispose of the rags properly. Keep vented batteries upright at all times.

The level of electrolyte in vented batteries must be checked on a regular basis. Use a hydrometer to check the quality of the electrolyte. In FLA batteries, distilled water may need to be added. Follow the manufacturer's instructions when adding water.

<table>
<tr><td>WARNING!</td><td>Do not add acid to an FLA battery. Adding acid will cause the battery to malfunction, and will void any warranties.</td></tr>
</table>

The deep-cycle batteries used in PV systems may have a life expectancy of up to 10 years, if well maintained and properly charged and discharged. Improper charging and discharging will damage the plates inside the batteries. Check the charging and discharging process at least annually.

3.5.0 Recommended Inverter Maintenance

Each inverter builder has a slightly different design, but most are very similar. Inverters must be cooled as they operate. Cooler air is pulled into the inverter, usually at the bottom because that is a more protected area. The heated air is usually exhausted near the top of the inverter. For this discussion, a Sunny Boy 3000US inverter is used as an example (*Figure 22*).

<table>
<tr><td>CAUTION</td><td>Do not blow compressed air into the air vent, fans, or handle covers of an inverter. Use a vacuum or soft bristle brush to clean these components.</td></tr>
</table>

Common maintenance activities on an inverter include the following:

- Cleaning the fans
- Cleaning the handle covers
- Operational checks

3.5.1 Cleaning the Fans

Before starting to clean or replace an inverter fan, open all DC and AC disconnects, and then wait five minutes for any residual voltages to dissipate. Follow the manufacturer's instructions to disconnect the inverter from both the DC and AC connections. After all the electrical connections are removed, verify that the fan is not moving. Next, carefully remove the plastic plate and filter that covers the fan (*Figure 23*).

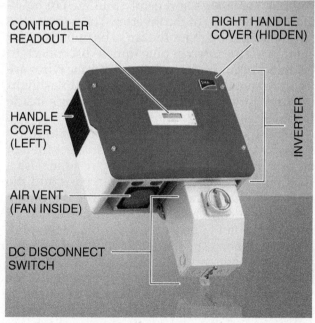

CONTROLLER READOUT

RIGHT HANDLE COVER (HIDDEN)

HANDLE COVER (LEFT)

INVERTER

AIR VENT (FAN INSIDE)

DC DISCONNECT SWITCH

105F22.EPS

Figure 22 Handle cover and air vent (and fan) locations on a typical inverter.

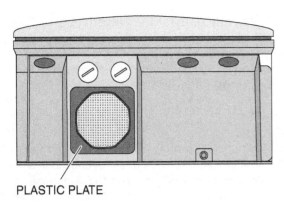

PLASTIC PLATE

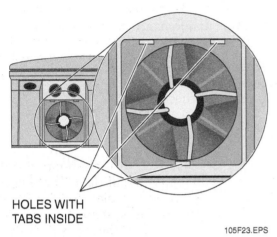

HOLES WITH
TABS INSIDE

105F23.EPS

Figure 23 Fan and cover location on a typical inverter.

In the inverter shown in *Figure 23*, plastic tabs hold the fan housing inside the inverter housing. Unhook the latches and gently pull the fan housing down and out of the inverter. The wires to the fan are long enough for the fan to be pulled out far enough for the internal plug (quick-disconnect) to be unlocked and disconnected. After the wires are free, remove the fan for cleaning. Use a flashlight to check for any buildup of dirt or debris inside the inverter housing.

While the fan is out, gently spin the fan to verify that it rotates with no resistance. Inspect the fan blades carefully for nicks and any residue. Use a soft brush or cloth to clean the fan parts and the filter and plastic housing that covered the fan. Do not use compressed air to clean the fan or filter. If the fan appears to be binding, replace it. If the fan blades appear damaged or unbalanced, it may be necessary to replace the fan assembly due to a damaged fan bushing or bearings. If the fan appears undamaged and is rotating correctly, replace it after it has been cleaned. Reconnect the fan wires, and then slide the fan housing back into the inverter housing. After the fan housing is back in place, replace the plastic housing and filter.

3.5.2 Cleaning the Handle Covers

Inverters have an exhaust fan and a vented opening to allow the heated air from inside the inverter to escape. The fan inside the inverter shown in *Figure 24* draws air in through the bottom and exhausts it out the two handle covers at the top right and left sides of the inverter. Any dirt that is in the air passing through the inverter accumulates on the fins or grill bars of the handle covers. The handle covers can be removed by placing fingers in the space near the top of each handle cover and gently pulling the cover up and out of its brackets.

After the covers have been removed, their fins and the filtering elements can be cleaned with a soft brush or cloth. If necessary, use a mild soapy water to remove any stubborn residue from the covers. Dry the covers after cleaning. Inspect them closely for cracks or breaks, and then carefully snap the covers back into place. Replace any damaged covers because they prevent insects and small animals from getting into the inverter housing.

3.5.3 Operational Checks

Like charge controllers, inverters are programmable and typically display system output data using an LED screen. Consult the manufacturer's instructions to interpret operating data and error codes (see *Appendix A* for an example of one manufacturer's error codes).

Inverters are designed to accumulate the incoming DC power and convert it into an AC supply that is compatible to the AC power from the grid. That means that the AC output of an inverter must be in phase with the grid power, and the

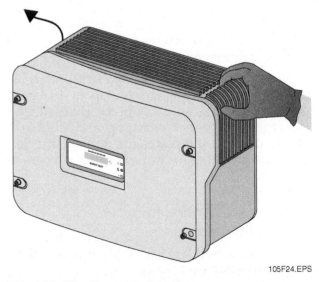

105F24.EPS

Figure 24 Handle cover removal.

level of power (voltage and current) must be high enough to meet the needs of the load. When checking an inverter, use a multimeter to verify the levels of DC power coming into the inverter, and the level of AC power being produced by the inverter. If possible, check the frequency of the outgoing power. Compare those numbers against the design plan for the inverter and determine if it is operating correctly. If any of those numbers are off, refer to the user manual for possible adjustments that can be made to the inverter. If the inverter cannot be adjusted to meet the minimum requirements, replace it.

4.0.0 TROUBLESHOOTING

Troubleshooting is a systematic process. The old trial and error method of troubleshooting is often a waste of time, and in some cases, a waste of parts. A troubleshooter needs to go into a situation with a plan and an understanding of the system being evaluated. That starts with examining the drawings and design plans originally laid out for the PV system being serviced.

Effective troubleshooting and problem resolution minimizes downtime and the unnecessary use of replacement parts. Troubleshooting starts with observations, either by the human senses or using test equipment. Effective troubleshooting procedures start at the inverter and work back upstream through a logical process of exclusion to determine the root cause of a problem. After the problem has been identified, the necessary corrective actions can be performed.

4.1.0 Alarms and Indicators

The programmable computers in charge controllers and inverters have digital readouts and flashing lights to indicate the operating status or alert the operator to a malfunction. Some PV system charge controllers and inverters are also tied through communication lines back to computers

that track trends in the equipment. Such tracking devices are a good source for historical data associated with the PV system.

While the charge controllers and inverters have the software to monitor their operations, the solar panels, combiners, DC disconnect switches, and wiring do not include alerting devices. If one of them starts to fail, the DC power into the charge controller or inverter will decrease. If the decrease is more than the charge controller or inverter can tolerate, the controller or inverter (or both) may be programmed to activate an alarm. In some cases, the equipment will simply shut down.

The combiner, and some disconnect switches, are equipped with circuit breakers or fuses, but most have no indicators to tell when a breaker has tripped or a fuse has blown. Without these indicators, PV maintenance personnel must rely on visual inspections and test equipment to determine when a component is not operating correctly.

4.2.0 Baseline Data

Any system check or troubleshooting effort must start with a baseline of data. That baseline data is the system design plan for the PV system. Always review the system design plan to become familiar with what the PV system should be doing. Make notes as to how much power the panels should be producing under ideal conditions. From that point on down, make notes as to what the DC and AC power levels should be at each test point. After the baseline data has been collected, start making electrical checks from the output of the final inverter to the output of the individual panels.

4.3.0 Breaking Down a PV System

Troubleshooting a PV system is no different than troubleshooting any other electrical system. All systems have at least one input, and they produce at least one output. If the input is there, but the output is not, mentally break the overall system down and start checking the inputs and outputs of each device within the system. If an inverter is not producing the desired level of AC power, check for a cause upstream. Is the inverter receiving the proper levels of DC power? If it is, the problem is in the inverter.

4.4.0 Inverters

While many systems use a charge controller with backup batteries, some PV systems take DC power directly from the array. If an inverter is equipped with a DC disconnect at its input, and that switch is opened, the inverter should not be receiving any DC power from any source. Since the inverter output is the point where the overall operation of the PV system is being evaluated, it is a good place to start the troubleshooting process.

One way to quickly evaluate an inverter is to supply it with a known level of DC power. If a stand-alone power supply (capable of generating the same level of DC power as the array) is available, the power supply's output can be temporarily hooked into the inverter's DC disconnect switch to simulate the DC power the inverter should be seeing. If such a power supply can be used, and the inverter functions correctly, the inverter can be eliminated as a problem component. In that case, the problem is somewhere upstream of the inverter. If such a power supply is used and the inverter still is not producing the correct AC power at its output, the problem is in the inverter. If the inverter cannot be adjusted so that it works correctly, replace it. After an inverter has been checked out and cleared as being functional, the temporary power supply can be removed and the inverter's DC disconnect switch can be closed to allow PV power to enter.

4.5.0 Charge Controllers

If the PV system is using a charge controller, the controller's job is to first charge the batteries until they are fully charged, and then send any excess DC power from the array on to the inverter. To simplify the process, temporarily disconnect the batteries from the charge controller. Doing so should make the controller send the DC power from the array directly to the inverter. If the charge controller is routing the array power directly to the inverter and the inverter output is not correct, either the array is not producing the expected level of DC power, or the charge controller is not working. If the charge controller cannot be adjusted to work correctly, replace it. If the inverter operates correctly on power supplied from the batteries and through the charge

controller, the charge controller is good. If the batteries and the charge controller are functional, but the charge controller and inverter are receiving no power from the array, check the combiner(s) and the panels.

4.6.0 Combiners and Panels

PV systems must have a DC disconnect switch on the output of the combiners. Opening the DC disconnect switch isolates the combiner from any downstream PV components. Opening the switch also prevents any accidental backfeed from any other source. With that switch open, the combiner should now only have DC power coming in from the solar panels.

Remember that combiners have either an individual fuse or a circuit breaker for each of the solar panel inputs. If all those fuses or circuit breakers are good and are allowing power to enter the combiner, the combiner should be producing a DC output equal to the sum of the multiple DC inputs from the panels. Use a multimeter to verify the inputs and outputs of the combiner. Remove fuses or open circuit breakers to isolate the individual inputs and take individual voltage readings from each panel input. Review the output voltage and current rating for each panel and then use a multimeter to see what the panel is actually producing. Make a note of each measurement. Add the multiple inputs to find the expected output of the combiner. After all the individual measurements are completed, close the circuit breakers or replace the fuses and then measure the summed output of the combiner. All combiners have a little loss, but the output should be very close to the desired level for the array. If the combiner does not show the correct output, it must be replaced.

If one of the panels supplying DC power to the combiner is below the minimal level set for the system, that panel may have to be removed and replaced. Before replacing any solar panel, make sure that it is clean and was not shaded during the test.

4.7.0 Manufacturer's Troubleshooting Data

The manufacturer's data normally includes troubleshooting information for the product (see *Appendix B*). This data typically lists common problems along with their possible causes and corrective actions. Using the manufacturer's troubleshooting information saves time and simplifies the process.

When encountering a problem that is not covered in the manufacturer's troubleshooting information, consult the equipment manufacturer or try the Internet for forums dealing with specific components. Start with a search of the manufacturer's home website. Such sites usually have links to any manual associated with a specific component, and may also have links to forums dealing with specific components or specific problems. The following are the web addresses for forums dealing with Outback and Sunny Boy (from SMA-America) equipment:

- www.outbackpower.com/forum
- http://forums.sma-america.com

4.8.0 Record Keeping

System troubleshooting begins with an understanding of what the system was designed to do when new, what it should be doing now, and what changes have occurred since it was installed. This process requires good maintenance records. After any work is done on a PV system, document all test results as well as any issues uncovered during the maintenance activities. If something minor was noticed during the preventive maintenance inspections, inform the customer and make a note in the records. For example, if the batteries are still operable but have begun to show signs of aging, let the customer know that they will soon require replacement. Also, be sure to indicate this in the system maintenance records. Small issues often result in system failure at a later date. Serious problems should always be resolved before leaving the site.

SUMMARY

This module has addressed the actions needed to maintain a basic PV system. When a good preventive maintenance program is in place for any system, the system will have less downtime and the equipment will last longer.

PV arrays are exposed to wind, weather, and other sources of damage. Because of their exposed conditions, additional care must be taken to keep the outside equipment clean and corrosion free. From a safety point of view, keeping the panels and their terminal boxes, combiners, and DC disconnect switches in good condition is critical since the panels are always producing electrical power (except at night). A broken wire could cause a fire, an electrical shock, or death.

The PV system components inside a structure are protected from the weather, but must also be cleaned, inspected, and tested. If batteries are used, they require regular testing and maintenance.

PV system maintenance begins by locating and reviewing the system records and the manufacturer's instructions for the equipment. Next, the system is inspected and all required maintenance performed. If the system is not producing the expected output, a systematic process of troubleshooting must be followed. Finally, records must be prepared of all system maintenance activities, test results, and repairs.

1. What category of test meters must be used on PV equipment?

   a. Cat I or II
   b. Cat II only
   c. Cat III / IV
   d. Cat V / VI

2. For a typical multimeter to read the frequency of a power line signal, the wire must be carrying at least _____.

   a. 1A
   b. 3A
   c. 4A
   d. 5A

3. The instrument used to measure solar irradiance is a _____.

   a. hydrometer
   b. refractometer
   c. pyrometer
   d. pyranometer

4. Before making any system performance checks, start with visual checks of the _____.

   a. panels and their mounting hardware
   b. combiners and wiring
   c. charge controller(s)
   d. inverter(s)

5. Who should have the historical records for a PV system?

   a. The installer
   b. The system owner
   c. The installation company
   d. The local building code office

6. In most cases, an inverter should have an output that is approximately _____.

   a. 65 percent of its input
   b. 75 percent of its input
   c. 85 percent of its input
   d. 95 percent of its input

7. When using a standard (non-MPPT) charge controller, the amount of power applied to the batteries is close to _____.

   a. 95 percent of the array output
   b. 90 percent of the array output
   c. 85 percent of the array output
   d. 80 percent of the array output

8. To clean PV system storage batteries, use _____.

   a. a mild acid such as vinegar
   b. plain water
   c. a mixture of water and salt
   d. a mixture of water and baking soda

9. What tool is used to clean inverter vents?

   a. Low-pressure water
   b. A vacuum cleaner
   c. Compressed air
   d. A wire brush

10. Which of the following is likely to include a digital readout for troubleshooting purposes?

    a. a combiner
    b. a DC disconnect
    c. an inverter
    d. solar panels

Cornerstone of Craftsmanship

Tony Vazquez
President/CEO
VisionQuest-Academy, Ocala, FL

Tony Vazquez came from a family of furniture-makers that emigrated from Cuba to the United States. Influenced by his father and grandfather, he entered the construction industry. It wasn't long before he and his father opened their own company, but the need to find good craftsmen led him to training. As the need for quality training became more apparent, he eventually opened his own construction training school. Tony enjoys sharing his love of building and encouraging others to excel in the industry that has made him successful.

How did you get started in the construction industry and who inspired you?
I was raised in a second-generation, construction/craft, working family. I was inspired as a boy by my father and grandfather back in Cuba, where they made furniture by hand. Later, here in the U.S., my father continued his craft as a construction worker and I, as a teenager, often went with him on weekends. After I left high school, my father and I started a construction company. The hard work that came with the job was outweighed by the gratification of seeing a well-built project that would last for many years. In 1990, while looking for trained construction workers, I ended up at the Miami Job Corps. I was impressed with the training going on in our industry. Later that year I volunteered my time to help with their program, and a few years later when the instructor retired, I was asked to take over. I returned to college and obtained my instructor/teacher certification in the industry and have enjoyed training others ever since.

What do you enjoy most about your job?
I still love to build. However, today I get the most satisfaction from teaching others the art of craftsmanship. I have taught at both the secondary (high school) and post-secondary (adult) levels for years. For decades the basics of construction have been taught at a few high schools (shop class) and a few post-secondary training programs scattered throughout the country. As our industry becomes more united in providing an industry-driven curriculum and nationally recognized training, it is revitalizing the art of construction craftsmanship. It has come during a much-needed time as our older workforce is preparing to retire and we have need of new blood and craftsmanship in our industry.

Do you think training and education are important in construction?
Yes, training and education are at a most influential time. Tough times today require specialized training. In today's way of building, each trade has become specialized, where those with the most training, experience, and credentials get the job. As our industry moves forward in creating nationally recognized credentials and certifications, so too have specialized training centers and schools at both secondary and post-secondary levels.

How important are NCCER credentials to your career?
Extremely important, in two distinct and different ways. First, by establishing standard trainer credentials, the industry ensures that training taking place across the country is similar and of the same quality. Second, completion points and credentials given to trainees are also similar and of the same quality across the country, thus making it possible to have national certification that is valid anywhere. At a personal level, having and acquiring multiple credentials allows me to excel. Reaching Master Trainer and Subject Matter Expert status with NCCER in the construction and green job trade has given me the opportunity to create my own construction training academy.

 Maintenance and Troubleshooting

How has training/construction impacted your life and your career?

Construction training has impacted my life since my teen years, both informally and formally. In the early years, it provided the technical background for the work I was performing. Years later, training became more formal in specialized areas. Currently, the combination of field work/training and formal classroom training has provided me with a good salary and lifestyle. Furthermore, it has provided me the opportunity to open VisionQuest-Academy, a full-service, post-secondary educational/training facility in Ocala, Florida. The academy is dedicated to providing cutting-edge construction and green job training, and courses are offered in English and Spanish. NCCER has been a key to this opportunity; with its support, many of our country's adults/working population now have the opportunity obtain national industry credentials.

Would you suggest construction as a career to others?

Yes, absolutely. As I tell my students, both youth and adults, it's never too late for training and education, either informal or formal. However, a well-planned training strategy is like a well-planned trip that includes a road map to your destination. NCCER has that covered for the construction and green job industry along with pathways to success. Ultimately, the construction and green job industry is especially rewarding in that, at the end of the day, you have something tangible to look back on.

How do you define craftsmanship?

Craftsmanship can be defined in a multitude of ways. For the most part, in the construction industry, craftsmanship is the final product of a job, task, or project completed within the quality standards set for that industry. However, I define it as follows: the skills, expertise, ability, and technique used by a person to shape, mold, transform, and convey an idea into products of value to others. A craftsman's level of work and performance can only be achieved by a combination of hard work, informal and formal training, time, and mostly a love of the craft.

Trade Terms Introduced in This Module

Refract: To alter the appearance of something by viewing or showing it through a different medium.

Specific gravity: The ratio of the weight of a substance to the weight of an equal volume of water at 60°F.

Appendix A

SUNNY BOY INVERTER ERROR MESSAGES

If a fault occurs, the Sunny Boy generates an error code according to the operating mode and the detected fault.

Error Type	Error Code	Description
Disturbance	Bfr-Srr	Communication between micro-controllers is failing. Contact SMA for assistance.
Warning	Derating	The inverter reduces the output power due to high internal temperature. Verify that the fans are operating normally and that the intake screens are clean. Check the intake vents for debris. Verify that there is adequate ventilation around the inverter. This condition may be normal in periods of high ambient temperatures (above 113°F).
Error	EarthCurMax-B	The earth current between PV+ and GND has exceeded the maximum limit. Check the PV array for ground faults.
Error	EarthCurMax-S	The earth current between PV+ and GND has exceeded the maximum limit. Check the PV array for ground faults.
Disturbance	EEPROM	Transition failure during reading or writing of data EEPROM. The data is not essential for safe operation and does not affect performance.
Error	EEPROM p	Data EEPROM defective, device is set to permanent disable due to the fact that the data loss affects important functions of the inverter. Contact SMA.
Disturbance	EeRestore	Internal failure.
Disturbance	Fac-Bfr, Fac-Srr	The AC grid frequency is exceeding the allowable range. The inverter assumes that the public grid is down and disconnects from the grid in order to avoid islanding. If the grid frequency is within the tolerable range and you still observe the failure message Fac-Bfr or Fac-Srr, check for possible intermittent connections. For additional assistance, contact SMA.
Warning	GFDI Fuse Open	The GFDI-Fuse is open or cleared. Check PV array for ground faults before replacing the fuse.
Disturbance	Grid-Timeout, Grid-Fault-S	The type of grid could not be detected (208/240).
Disturbance	Imax	Overcurrent on the AC side. This failure code is indicated in case the current to the AC grid exceeds the specification. This may happen in case of harmful interference on the grid. If you observe Imax often, check the grid. Contact SMA for assistance.
Disturbance	K1-Close	Relay test failed. Contact SMA for assistance.
Error	K1-Open, K2-Open	Relay test failed. Contact SMA for assistance.
Disturbance	MSD-FAC, MSD Idif	Internal measurement comparison error: The internal BFS and SRR processors are measuring different values. Contact SMA for assistance.
Error	MSD-VAC	Internal measurement comparison error: The internal BFS and SRR processors are measuring different values. Contact SMA for assistance.
Disturbance	OFFSET	Grid monitoring self-test failed. Contact SMA for assistance.
Error	ROM	The internal test of the Sunny Boy control system firmware failed. Contact SMA in case you observe this failure often.
Disturbance	Shut-Down	Internal overcurrent continuous. Contact SMA for assistance.
Disturbance	Vac-Bfr, Vac-Srr	The AC grid voltage is exceeding the allowable range. Vac can also result from a disconnected grid or a disconnected AC cable. The inverter assumes that the public grid is down and disconnects from the grid in order to avoid islanding. If the grid voltage is within the tolerable range and you still observe the failure message Vac-Bfr or Vac-Srr, contact SMA for assistance.

Error Type	Error Code	Description
Disturbance	VacL1-Bfr, VacL2-Bfr, VacL1-Srr, VacL2-Srr	Voltage is too high or too low on the indicated leg.
Disturbance	VpvMax !PV Overvoltage! !Disconnect DC!	DC input voltage above the maximum tolerable limit. Disconnect DC immediately!
Disturbance	Watchdog	Watchdog for operation control triggered. Contact SMA for assistance.
Disturbance	XFMR_TEMP_F	High transformer temperature. The inverter will remain stopped until the transformer has cooled.
Warning	XFMR_TEMP_W	High transformer temperature is gone. The inverter starts working and shows the failure XFMR_TEMP_W. Verify that the fans are operating normally and that the intake screens are clean. Check the intake vents for debris. Verify that there is adequate ventilation around the inverter. This condition may be normal in periods of high ambient temperatures (above 113°F).

Appendix B

OUTBACK CHARGE CONTROLLER TROUBLESHOOTING GUIDE

Charge Controller Does Not Boot/Power-Up (Blank LCD)

- Check the battery connection and polarity. Reverse polarity or an improper connection will cause power-up issues.
- Check the battery breaker. Ensure that the battery breaker is sized appropriately.
- A battery voltage below 10.5VDC may not power up the charge controller (measure the battery side of wire lugs).
- If the charge controller still does not power up, call the factory for additional support.

Charge Controller is Always SLEEPING

- If the battery voltage is at or above the ABSORB voltage set point (compensated ABSORB voltage), the charge controller will not wake up.
- The PV voltage has to be at least two volts greater than the battery voltage for the initial wakeup.
- Check the PV array breaker (or fuse).
- Confirm the PV array breaker (or fuse) is sized appropriately.
- Which state (in MISC Menu) is it at? Is it transitioning between 00 and 01? Is it in GT mode and connected to a MATE? GT mode is only applicable with a HUB 4 or HUB 10 installations with a grid-tie compatible MATE.
- Does the PV array voltage on the display rise with the PV breaker OFF, but reads 000 with the PV breaker on? If so, the PV array polarity connection on the charge controller may be reversed or the PV lines could be shorted.
- Does the PV voltage still read 000 with the PV breaker off after a minute? Call the factory for support.
- Have you checked the short circuit current of the PV array? Use a multimeter to determine if a short circuit current is detected. The short circuit current test will not harm the array.

Charge Controller Not Producing Expected Power

- Clouds, partial shading, or dirty panels can cause poor performance. The lower current limit set point in the Charger menu will yield a loss of power or poor performance symptoms.
- Are the batteries charged? Is the charge controller in the ABSORB or FLOAT stage? If either case is true, the charge controller will produce enough power to regulate the voltage at the ABSORB or FLOAT set point voltage, therefore, requiring less power in these modes.
- What is the short circuit current of the PV array? Use a multimeter to determine if the short circuit current is as expected. There might be a loose PV array connection.
- If the PV array voltage is close to the battery voltage, the panels could be warm/hot causing the maximum power point to be at or lower than the battery voltage.
- Is it in U-Pick mode?

Charge Controller is Not Equalizing

- Has the EQ cycle been initiated? In the EQ Menu, press START to begin the process. When the EQ cycle has been initiated, EQ-MPPT will be displayed.
- The EQ cycle has been initiated, but the battery is not equalizing. The EQ cycle will begin when the target EQ set point voltage has been reached. A small array or cloudy weather will delay the EQ cycle. Accordingly, running too many AC and/or DC loads will also delay the EQ cycle.
- An EQ set point that is too high relative to the battery voltage will delay the EQ cycle.
- If the PV array voltage is close to the battery voltage, the panels could be warm/hot, causing the maximum power point to be at or lower than the battery voltage, which can delay the EQ cycle.

Charge Controller Battery Temperature Compensated Voltage

- Only the Outback RTS (remote temperature sensor) can be used with the charge controller.
- The battery voltage can rise above the ABSORB and FLOAT voltage set points if the battery temperature is < 77°F or fall below the ABSORB and FLOAT voltage if the battery temperature is > 77°F.
- If the charge controller shows BatTmpErr on the STATUS screen, the RTS is faulty or damaged. Disconnect the RTS from the RTS jack to resume normal operation.

Charge Controller Internal Fan

- The internal fan will only run when the internal temperature has reached approximately 112°F.
- The fan will continue running until the internal temperature is less than 104°F.

Charge Controller is Beeping

- When the charge controller is in Extended Play mode, the array is very hot, and the maximum power point is close to the battery voltage, or the nominal PV voltage is higher than the nominal battery voltage, beeping can occur. To disable the Extended Play feature, go to the MAIN Menu and press and hold the #1 soft key until the charge controller's software version appears on the screen. Continue pressing the #1 soft key and press the #3 soft key at the same time until X Off displays on the screen.
- To reactivate Extended Play, repeat these steps and hold the #3 soft key until X On displays. Extended Play is meant to optimize the performance of a hot array, but isn't critical to efficient charge controller operation.

Additional Resources

This module presents thorough resources for task training. The following resource material is suggested for further study.

Canadian Solar website:
    www.canadian-solar.com.

Florida Solar Energy Center website:
    www.fsec.ucf.edu.

Outback Power Systems website:
    www.outbackpower.com.

Solar panel terms: www.osha.gov.

Solar Source Institute website:
    www.solarsource.net.

Surrette/Rolls Battery website:
    www.surrette.com.

Figure Credits

NCCER CURRICULA — USER UPDATE

NCCER makes every effort to keep its textbooks up-to-date and free of technical errors. We appreciate your help in this process. If you find an error, a typographical mistake, or an inaccuracy in NCCER's curricula, please fill out this form (or a photocopy), or complete the online form at **www.nccer.org/olf**. Be sure to include the exact module ID number, page number, a detailed description, and your recommended correction. Your input will be brought to the attention of the Authoring Team. Thank you for your assistance.

Instructors – If you have an idea for improving this textbook, or have found that additional materials were necessary to teach this module effectively, please let us know so that we may present your suggestions to the Authoring Team.

NCCER Product Development and Revision
13614 Progress Blvd., Alachua, FL 32615

Email: curriculum@nccer.org
Online: www.nccer.org/olf

❑ Trainee Guide   ❑ Lesson Plans   ❑ Exam   ❑ PowerPoints   Other _____

Craft / Level: _____ Copyright Date: _____

Module ID Number / Title: _____

Section Number(s): _____

Description: _____

_____

_____

_____

_____

Recommended Correction: _____

_____

_____

_____

_____

Your Name: _____

Address: _____

_____

Email: _____ Phone: _____

Glossary

Absorption stage: The second stage of battery charging, where the voltage remains constant and current is gradually reduced as resistance in the circuit increases. This stage continues until a full charge condition is sensed. During this stage, the charging voltage is typically highest, from roughly 14V to 15.5V.

Air mass: The thickness of the atmosphere that solar radiation must pass through to reach the Earth.

Altitude: The angle at which the sun is hitting the array.

Ambient temperature: The air temperature of an environment.

Amorphous: A low-efficiency type of photovoltaic cell characterized by its ability to be used in flexible forms. Also known as thin film.

Array: A complete PV power-generating system including panels, inverter, batteries and charge controller (if used), support system, and wiring.

Autonomy: The number of days a fully charged battery system can supply power to loads without recharging.

Azimuth: For a fixed PV array, the azimuth angle is the angle clockwise from true north that the PV array faces.

Backfeed: When current flows into the grid.

Balance of system (BOS): The panel support system, wiring, disconnects, and grounding system that are installed to support a PV array.

Battery bank: A collection of batteries electrically connected in parallel and series combinations to generate the desired voltage and current capacity needed.

Brownout: A temporary decrease in grid output voltage typically caused by peak load demands.

Building-integrated photovoltaics (BIPV): A PV system built into the structure as a replacement for a building component such as roofing.

Bulk charge stage: The initial stage of three-stage battery charging, where the maximum amount of current is delivered to the battery until it has reached 80 to 90 percent of its possible charge capacity. Voltage generally ranges from 10.5V to 15V through this stage.

Bulk voltage setting: The voltage setpoint for a battery charge controller to perceive as roughly 80 to 90 percent of the battery's fully charged state, initiating the absorption stage.

Bypass diode: A diode used to direct current around a panel rather than through it. Bypass diodes are typically used to overcome partial shading.

Charge controller: A device used to regulate the charging and discharging of the battery system to prevent overcharge and excess discharge.

Combiner box: A junction box used to connect strings of solar panels to create a larger array, and to provide a convenient array disconnect point.

Concentrator: A device that maximizes the collection of solar energy by using mirrors or lenses to focus light onto specially designed cells.

Days of autonomy: The number of days that a solar PV system can continue to meet an electrical load requirement without additional input from the solar array.

Declination: The angle between the equator and the rays of the sun.

Depth of Discharge (DOD): A measure of the amount of charge removed from a battery system.

Diversion controller: A style of battery charge controller which diverts PV array capacity to another DC load whenever the battery charging process is inactive.

Doped: A material to which specific impurities have been added to produce a positive or negative charge.

Dual-axis tracking: An array mounting system designed to adjust both the horizontal and vertical axes of a panel to precisely follow the movement of the sun.

Electrochemical solar cells: A type of PV cell that replaces silicon with a light-sensitive dye that absorbs light and produces current.

Elevation: A measure of a location's relative height in reference to sea level.

Equalization stage: A special charging cycle used intermittently on flooded batteries to rejuvenate and circulate the electrolyte through gassing. Higher than normal charge voltages are applied for a specified period of time, creating a bubbling activity inside to redistribute the electrolyte.

Equipment grounding: Connecting the chassis or frame of an appliance or unitary piece to ground. This ground should connect all components of an installation which do not conduct current and should be uninterrupted.

Float stage: The third stage of three-stage battery charging, where charging voltage is reduced to maintain the fully charged condition of the battery. This is often referred to as at trickle charge or maintenance charge.

Fuel cell: A device that harnesses the energy produced by a chemical reaction between hydrogen and oxygen to produce direct current.

Grid-connected system: A PV system that operates in parallel with the utility grid and provides supplemental power to the building or residence. Since they are tied to the utility, they only operate when grid power is available. Also known as a grid-tied system.

Grid-interactive system: A PV system that supplies supplemental power and can also function independently through the use of a battery bank that can supply power during outages and after sundown.

Grid-tied system: See *grid-connected system*.

Ground fault protection: The incorporation of devices which monitor the passage of current in a hot conductor and compares it to that passing through the neutral conductor. An imbalance indicates current flow to ground and the ground fault protection device actuates to open the circuit.

Heat fade: A condition in which a PV system operates inefficiently during periods of high heat. Heat fade is usually caused by poor connections or undersized wiring.

Hybrid system: A grid-interactive system used with other energy sources, such as wind turbines or generators.

Insolation: The equivalent number of hours per day when solar irradiance averages 1,000 W/m². Also known as peak sun hours.

Interconnection agreement: An agreement that specifies the terms and conditions of power buying and selling between the utility and the PV system owner or operator, and outlines the technical requirements of the PV system connection interface.

Inverter: A device used to convert direct current to alternating current.

Irradiance: A measure of radiation density at a specific location.

Latitude: A method of determining a location on the Earth in reference to the equator.

Low-voltage disconnect (LVD): The recommended lowest level of battery discharge, measured in volts, and specified by the manufacturer for a given battery type and construction. This value is used as a control setpoint to disconnect the battery from the load to prevent its permanent damage.

Magnetic declination: The angle between magnetic north, as indicated by a compass needle, and true north (the North Pole).

Maximum power point tracking (MPPT): A battery charge controller that provides precise charge/discharge control over a wide range of temperatures.

Module: A PV system component consisting of numerous electrically and mechanically connected PV cells encased in a protective glass or laminate frame. Also known as a PV panel.

Monocrystalline: A type of PV cell formed using thin slices of a single crystal and characterized by its high efficiency.

Net metering: A method of measuring power used from the grid against PV power put into the grid.

North American Board of Certified Energy Practitioners (NABCEP): A volunteer board of renewable energy system professionals that provides standardized testing and certification for PV system installers.

Off-grid system: A PV system typically used to provide power in remote areas. Off-grid systems use batteries for energy storage as well as battery-based inverter systems. Also known as a stand-alone system.

Peak sun hours: See *insolation*.

Photovoltaic (PV) cell: A semiconductor device that converts sunlight into direct current.

Pitch: The ratio of the rise of a roof to its span expressed as a simple fraction.

Polycrystalline: A type of PV cell formed by pouring liquid silicon into blocks and then slicing it into wafers. This creates non-uniform crystals with a flaked appearance that have a lower efficiency than monocrystalline cells.

Pulse width-modulated (PWM): A control that uses a rapid switching method to simulate a waveform and provide smooth power.

Refract: To alter the appearance of something by viewing or showing it through a different medium.

Reverse bias: A PV cell or panel operating at a negative voltage, typically due to shading.

Sea level: A measure of the average height of the ocean's surface between low and high tide. Sea level is used as a reference for all other elevations on Earth.

Semiconductor: A material that exhibits the properties of both a conductor and an insulator.

Shunt controller: A simple form of battery charge controller which shorts the PV array charging power to a false load when the charging process is complete, creating heat.

Single-axis tracking: An array mounting system designed to adjust either the horizontal or the vertical axis of a panel to follow the movement of the sun.

Single-stage controller: A battery charge controller which opens the charging circuit from the PV array once charging is complete. Most incorporate other features, such as float stages, once this level of controller design is reached.

Solar combiner: An application-specific junction box where multiple solar panels are connected together, combining their output together before being routed to the charge controller or inverter.

Specific gravity: The ratio of the weight of a substance to the weight of an equal volume of water at 60°F.

Spectral distribution: The distortion of light through Earth's atmosphere.

Standalone system: See *off-grid system*.

Standard Test Conditions (STC): Standardized panel ratings based on a specific operating temperature, solar irradiance, and air mass.

Sulfation: A chemical process inside a battery which causes large crystals of lead sulfate to form on the plates, instead of desirable minute crystals. This process generally occurs in batteries which experience poor charging cycles or remain discharged.

Sun path: The sun's altitude and azimuth at various times of year for a specific location or latitude band.

System grounding: Connecting one conductor of a two-conductor system to ground.

Thin film: See *amorphous*.

Tilt angle: The position of a panel or array in reference to horizontal. Often set to match local latitude or in higher-efficiency systems, the tilt angle may be adjusted by season or throughout the day.

Utility-scale solar generating system: Large solar farms designed to produce power in quantities large enough to operate a small city.

Watt-hours (Wh): A unit of energy typically used for metering.

Index